THE ARTHUR YOUNG GUIDE TO
WATER and WASTEWATER FINANCE and PRICING

GEORGE A. RAFTELIS

 LEWIS PUBLISHERS

Library of Congress Cataloging-in-Publication Data

Raftelis, George A.
  The Arthur Young guide to water and wastewater finance and pricing
  George A. Raftelis.
    p.   cm.
  Includes index.
  ISBN 0-87371-181-5
  1. Water-supply — Finance.   2. Water-supply — Rates.   3. Sewage
disposal — Finance.   4. Sewage disposal — Rates.   I. Arthur Young &
Company. II. Title.
HD4456.R34   1989   89-2502
628.1′068 — dc19   CIP

Second Printing 1989

LEWIS PUBLISHERS, INC.
121 South Main Street, Chelsea, Michigan 48118

PRINTED IN THE UNITED STATES OF AMERICA

To my beloved mother, Josephine,
and my devoted wife, Eva

PREFACE

As Director of Arthur Young's National Environmental Consulting Practice, I have had the opportunity to work with over 200 water and wastewater utilities across the country. Most of my assistance to these utilities has been in the area of financial planning and pricing.

There has long been a need for a document that discusses, in an understandable manner, some of the common themes and trends across the country in water and wastewater financial planning and pricing. Numerous approaches to water and wastewater financial planning and pricing exist, and it is important to evaluate these different approaches and tailor a financial plan and pricing structure to address the unique requirements of each utility. My purpose in writing this book is to share with the industry my experience in effective financial planning and pricing techniques. Specifically, the book provides a framework from which utility professionals can work in addressing their financial planning and pricing objectives. This book is a management tool as well as a technical guide to effective financial planning and rate-setting practices.

This book is intended for anyone dealing with the financial and pricing concerns of water and wastewater utilities. I have learned much from this audience over the last 15 years and use this book as a means to share important principles and practices related to financing and pricing. This book is a dynamic document that will be updated as financial planning and pricing techniques change to address changing utility environments.

ACKNOWLEDGMENTS

I have many people to thank for input into this book. Many hours have been spent by my administrative staff (Pat Turpin, Shirley Harvey, Marlond Meadows, and others) in going through the numerous iterations necessary to develop a quality product. Christie Stewart and Michele Miller have made major contributions to the book by carefully editing the text. In addition, Christie has devoted many hours to coordinating with Lewis Publishers during the final stages of production. Sue Nichols provided valuable assistance in preparing graphics for the book. Attorneys, investment bankers, rate technicians, financial specialists, and other professionals have provided direct input on very specific issues discussed in the book. Professionals in Arthur Young's Environmental Consulting Practice have provided positive input in improving the book. Special thanks go to my Technical Evaluation Committee for evaluating the technical merits of the book and providing recommendations for improving technical quality. Most of all, I thank my clients for providing me with the opportunity to understand their concerns and develop solutions to their financing and pricing needs. Without all of these parties, this book could not have been written.

The author, **George Raftelis**, is a partner with Arthur Young & Company and resides in Charlotte, North Carolina. He is Director of Arthur Young's National Environmental Consulting Practice, specializing in providing financial, management, accounting, and public policy consulting services to the environmental industry. His clients include public and private sector organizations involved in the environmental disciplines of water, wastewater, solid waste, hazardous waste, air pollution, noise pollution, and stormwater.

Mr. Raftelis has approximately 15 years of consulting experience to the environmental industry. He has provided environmental consulting assistance to over 150 communities, both large and small, across the country. His experience includes the development of impact fees (capital recovery charges), water and wastewater cost-of-service and rate structure studies, the development of water and wastewater financial plans, valuation and acquisition feasibility analysis, litigation support services related to water and wastewater pricing and finance, and assisting utilities in selecting the appropriate pricing structure and financing plan.

The author received a BS in mathematics with a minor in economics from Eckerd College in St. Petersburg, Florida, in 1969. Following four years of service as an officer in the Military Intelligence Branch of the United States Army, he attended the Fuqua School of Business at Duke University and received an MBA in 1975. He then joined Arthur Young & Company in Charlotte, where he established Arthur Young's Environmental Consulting Practice.

He has published technical articles in periodicals such as *American City & County*, *American Public Works*, *American Water Works Journal*, and other leading trade journals. In addition, he has developed and conducted numerous technical workshops for professional associations and environmental agencies such as the American Water Works Association, the Water Pollution Control Federation, the Environmental Protection Agency, the Water Resources Research Institute, the Government Finance Officers Association, the American Society of Public Administrators, the National League of Cities, the U.S. Conference of Mayors, and the International City Management Association.

Mr. Raftelis is active in the American Water Works Association, the Water Pollution Control Federation, and the American Institute of Certified Public Accountants. For the American Water Works Association, he participates on the Committee on Rates.

TECHNICAL EVALUATION COMMITTEE

 **Edward G. Blundon** is the Assistant Water and Wastewater Director–Administration for the City of Phoenix, Arizona. He has had over 18 years of public management experience in budget, accounting, rate, and financial areas. In 1982, Mr. Blundon received the Productivity Innovator award, and in 1985 he received the Excellence in Management award, both from the City of Phoenix. Mr. Blundon is a member of the American Water Works Association, the American Public Works Association, and the International City Management Association. He has a BA in economics from Bethany College in Bethany, West Virginia, and has completed additional Masters studies in public administration at California State University in Los Angeles.

Deborah T. Broome currently serves as Personnel Director for the City of Louisville, Colorado. Prior to this position, she was a Financial Analyst for the City of Boulder, Colorado Public Works/Utilities Department, where she coordinated all budgetary and financial aspects of water, wastewater, flood control, and hydroelectric utilities. Prior to joining the City of Boulder, Ms. Broome held several positions with Alachua County, Gainesville, Florida. Ms. Broome has a BA and an MA in public administration from the University of Florida. She has also served in a number of professional organizations, including the Committee on Rates of the American Water Works Association, the Water Pollution Control Federation, and the Executive Board of the American Society for Public Administration.

Jeffrey C. Carey joined Merrill Lynch in 1979 and currently serves as senior banker for numerous issuers in Merrill Lynch's northeast region. He has worked on a wide range of negotiated financing — bonds, notes, variable rate demand obligations, commercial paper, and certificates of participation. Mr. Carey received a BA in regional planning, with honors, from Dickinson College and earned a Master's degree in city and regional planning, with distinction, in public finance from Harvard University. Mr. Carey is registered with the New York Stock Exchange and the National Association of Securities Dealers.

Sandra M. Doyle is currently Manager of the Finance Division for the Public Works Department of the City of Albuquerque, New Mexico. In her position, she is responsible for the development, implementation, and monitoring of water and wastewater operating and capital budgets. She supervises, directs, coordinates, and prepares financial activities such as purchasing, cash flow analysis, grants administration, budget management, rate analysis, and planning. Prior to her work at the City of Albuquerque, Ms. Doyle was a staff accountant with the international accounting firm of KPMG Peat Marwick (formerly Peat Marwick and Mitchell). She has a Bachelor of University Studies and an MBA with an emphasis in accounting from the University of New Mexico.

 **James W. Fagan** is a principal with JWF Associates, a consulting firm specializing in environmental management and finance. He is experienced in government utility finance and management, technical aspects of water and wastewater utility operations, administrative and regulatory proceedings, and environmental program development. He has either authored or coauthored several authoritative publications dealing with government utility management and finance. Before establishing JWF Associates, Mr. Fagan served as a management consultant with the international accounting firm of KPMG Peat Marwick (formerly Peat Marwick and Mitchell), as a financial analyst with Shearson Hammill & Company, and as an engineer for Hazen & Sawyer. Mr. Fagan has a Bachelor's degree in civil engineering and a Master's degree in engineering from Manhattan College, as well as a Juris Doctorate degree from Fordham University.

 **Roger D. Feldman** is an environmental and finance attorney with Nixon, Hargrave, Devans & Doyle. He has an undergraduate degree from Brown University, an MBA from the Harvard Business School, and a law degree from Yale Law School. Mr. Feldman has been listed in *Who's Who in America* and *Who's Who in American Law*. He was also a member of Phi Beta Kappa fraternity. Mr. Feldman's professional memberships include the American Bar Association (chairman, Environmental Values Committee); the Association of Energy Engineers (board member, Cogeneration Institute); and the Privatization Council (vice chairman). He has authored articles in the *Public Utilities Fortnightly*, *Privatization Review*, *Waste Age*, and *Cogeneration Journal* and is on the editorial board of *Waste Tech News*. With Nixon, Hargrave, Devans & Doyle, Mr. Feldman has provided legal counsel to jurisdictions, developers, and financial institutions in the area of infrastructure financing.

George W. Johnstone is Senior Vice President of the American Water Works Service Company and Vice President of its parent, the American Water Works Company. He joined the American System in 1966, after working with the Pennsylvania Department of Health and a consulting engineering firm, and has been employed there since then. Mr. Johnstone graduated from Pennsylvania State University in 1960 with a degree in sanitary engineering. He is a registered professional engineer in seven states and a member of the American Society of Civil Engineers. He is also a member of the Water Pollution Control Federation and of the American Water Works Association, where he served as chairman of the Committee on Rates.

David Mackenzie is currently a Vice President with R-C Capital, the project finance subsidiary of Research-Cottrell, Inc. (the parent of Metcalf & Eddy). He holds a BS degree from Boston University and a Master's degree in public administration from the University of New Hampshire. Mr. Mackenzie has spent over 20 years in project development, including public finance, investment banking, financial advisory services, and credit enhancement consultation. He was a major contributor to Arthur Young's *The Privatization Book*, covering privatization analyses, procurement, and financing.

Richard M. Marvin currently manages public finance for Wachovia Bank in Winston-Salem, North Carolina. He received an Associate Degree from Brevard College, an undergraduate degree from the University of North Carolina at Chapel Hill, and a Master's degree in public administration from American University in Washington, DC. His memberships include the International City Management Association, the Government Finance Officers Association, and the Bank Capital Markets Association. Mr. Marvin has published articles in *Government Finance Review* and *Public Manager*. Prior to joining Wachovia Bank, he was in city management at Myrtle Beach, South Carolina; West Palm Beach, Florida; and Streamwood, Illinois.

W. Carey Odom is Treasurer of the City of Charlotte, North Carolina. He has had approximately 20 years of financial and accounting experience and has served on many projects related to bond financing, water and wastewater pricing, and other government utility issues. He is a graduate of Pembroke State University, North Carolina, and a member of the Government Finance Officers Association and the North Carolina Public Finance Officers Association.

Myron A. Olstein is a principal with the international accounting firm of KPMG Peat Marwick. He is a Certified Management Consultant and a panelist of the American Arbitration Association, and a member of the Task Force on Financial Reporting by Public Authorities for the Government Accounting Standards Board. His professional memberships include the Institute of Management Consultants, the Government Finance Officers Association, and the American Water Works Association. Mr. Olstein has published articles in the *Privatization Journal*, *Municipal Finance*, and *Governmental Finance Review*. He was also a contributor to the book *Financing the Commonwealth*. He graduated with a BA and BS in chemical engineering from Lehigh University in Bethlehem, Pennsylvania, and received a Master's degree from George Washington University in Washington, DC. Mr. Olstein has more than 15 years of experience in providing financial and pricing assistance to municipal utilities.

Sam F. Rhodes is a partner with the international accounting firm of Touche Ross & Company. He is a Certified Public Accountant and a Certified Management Consultant. Mr. Rhodes' memberships include the American Water Works Association, the American Institute of Certified Public Accountants, and the Institute of Management Consulting. Mr. Rhodes has performed numerous water and wastewater cost-of-service/rate design studies, and has provide testimony in several regulated jurisdictions and in federal and state courts. Mr. Rhodes received a BS in agricultural economics from Texas A&M University.

 **Douglas P. Wendel** currently serves as Executive Director of the Grand Strand Water & Sewer Authority in Horry County, South Carolina. Prior to joining the Authority, Mr. Wendel served in a number of administrative and management positions for Anne Arundel County, Maryland; was the City Manager of North Myrtle Beach, South Carolina; served as County Administrator for Horry County, South Carolina; and held a number of administrative positions for the U.S. Congress. He received an undergraduate degree and a Master's degree in public administration from the University of Maryland. Mr. Wendel has also served in a number of professional and civic organizations, including the American Society for Public Administration, the Government Finance Officers Association, the National Association of Counties, the International City Management Association, the Congressional Administrative Assistant Association, and the Rotary Club. For the American Water Works Association, Mr. Wendel serves as a member of their Utility Council. During his career, Mr. Wendel has developed extensive expertise in the area of water and wastewater management and finance.

Jean Williams is a water resources consultant in private practice. She has a degree in geology from the University of Texas at Austin. For approximately six years she was a Vice President with Camp Dresser & McKee Inc., where she provided technical services for projects involving regional and state water resources planning and management. Ms. Williams was a technical consultant to the City of San Antonio and Edwards Underground Water District on an institutional and financing study for the area. In addition, she was consultant to Hays County in the development of a long-range water resource management study.

Christopher P. N. Woodcock is an engineer with Camp Dresser & McKee Inc. in Boston. He has performed nearly 200 water and sewer rate and financing studies, many of which have been for major New England cities. He was elected public works commissioner for the town of Wayland, Massachusetts, responsible for adopting local government utility rates. He has a BA in economics and a BS in environmental engineering from Tufts University. His professional memberships include the American Water Works Association, the Water Pollution Control Federation, and the New England Water Works Association. He has published articles in *Water Engineering and Management* and *Journal of the New England Water Works Association.*

CONTENTS

APPENDIXES

FIGURES

TABLES

1   INTRODUCTION TO WATER AND WASTEWATER FINANCING AND PRICING

1   INTRODUCTION TO WATER AND WASTEWATER FINANCING AND PRICING

Few things are more important in our country than providing high-quality potable water and insuring that wastewater is properly treated and returned to the environment. Most Americans take for granted the tens of thousands of people who are employed daily in addressing America's water and wastewater needs. Utility managers are responsible for making sure that proper water and wastewater services are provided to residents, businesses, industries, and other customers within a community. Water and wastewater employees work in utility operations and administrative support areas. Utility governing bodies must ensure that customer interests are understood and that utilities properly address public concerns. Regulatory agencies at the federal and state levels are charged with the responsibility of ensuring that utilities are protecting the public by appropriately addressing water quality regulations. The investment community has the responsibility to generate that front-end capital to finance major water and wastewater facilities. Environmental attorneys protect the legal rights of utilities, utility customers, and others that might be affected by water and wastewater operations and services. Consultants assist utilities in addressing financial, economic, engineering, and other professional requirements. Clearly, a broad spectrum of Americans plays a part in ensuring that appropriate water and wastewater services are provided to the public.

The operations and management of water and wastewater have changed significantly over the last 20 years. At one time, providing water and wastewater operation services was reasonably simple. Water and wastewater pollution did not appear to be a high priority concern of most Americans. In the late '50s, the public became increasingly concerned about protecting the environment. Through the efforts of many concerned Americans, the U.S. Environmental Protection Agency (EPA) was formed in the early 1960s to address environmental quality issues. Since that time, there has been an increased focus on providing high-quality potable water and ensuring that wastewater is treated to appropriate levels to protect the quality of streams and waterways. In 1972, one of the most comprehensive water quality laws was passed. The law, Public Law 92–500 or the Water Pollution Control Act Amendments of 1972, was targeted toward ensuring water quality standards throughout America and mandated the largest public works program in history. All water and wastewater utilities were affected by this law and billions of dollars in grant funds became available to governmental utilities for constructing facilities to address water quality standards. In addition, the Safe Drinking Water Act Amendments mandate significant requirements applying to thousands of public water purveyors across the country.

As a result of increased focus on environmental quality, the water and wastewater industries have become more sophisticated. More technically advanced water and wastewater treatment facilities have been constructed to provide high-quality potable water and to ascertain that wastewater is treated to appropriate levels. More sophisticated equipment has been developed to provide better support to operations. Highly advanced information systems ensure that proper management and accounting information is provided to management, operators, and others involved in providing water and wastewater services. More highly trained operators are necessary to run more advanced equipment and utility facilities. Highly educated and experienced managers are required to deal with the complex management, financial, engineering, and political issues. As a result, many people have decided to dedicate their careers to this expanding industry.

A major challenge confronting the water and wastewater industries is acquiring adequate funds to finance capital equipment and facilities and implementing appropriate pricing structures to ensure self-sufficiency of the utility. Capital financing is important because it is necessary to make sure that appropriate facilities will be constructed to address environmental regulations and meet the service needs of the customers. The financing vehicle that is used by the utility and the timing of the financing is crucial in ensuring that each generation of water and wastewater customers is appropriately paying for facilities that they need, and not inappropriately financing facilities for other generations of customers. It is a major goal of an effective financial plan to "match" economic impact on customers with benefits received by these customers.

Water and wastewater pricing, typically referred to as user charges or rates, provides a direct form of communication with the customer. Rates generate the major source of utility revenues and define the customer's obligation to participate in the costs of operating and maintaining the utility. When rates become excessive, the public provides feedback to the utility about paying higher charges.

Almost everyone in our society is affected by water and wastewater financing and pricing. Residential customers are affected in that water and wastewater connection charges and rates have to be factored into their personal finances and budgets. Commercial establishments have to consider water and wastewater utility costs in pricing their goods and services. Manufacturing companies closely scrutinize water and wastewater rates, as they are under pressure to provide manufactured goods at the lowest possible prices in order to compete with other companies. As water and wastewater quality has improved, the costs of providing water and wastewater services have increased. As a result, charges for these services have become a more significant part of the budget of each water and wastewater customer.

With additional scrutiny by customers, one of a water and wastewater utility's major objectives is to provide service at the most affordable price. At the same time, the utility must see that it complies with appropriate government regulations and that adequate funds are raised to maintain the utility on a financially self-sufficient basis. In addition, the utility must be sensitive to the cost of providing service, and allocate this burden equitably to users based on the cost of providing service to these users.

This book is organized into two parts. Part I (Chapters 2 through 6) deals with the financial planning challenges of a utility. Specifically discussed is an appropriate process for developing a financial plan with effective financing vehicles. Chapter 2 provides an overview of this process and discusses major components of the process. Chapter 3 discusses short-term financing techniques such as fixed rate demand notes, tax-exempt commercial paper, and variable rate demand notes. Also discussed in Chapter 3 are credit facilities for short-term financing and long-term financing methods such as general obligation bonds, revenue bonds, and double-barrel bonds. In addition, Chapter 3 discusses innovative techniques that have become popular over the last several years in the long-term and short-term bond market (variable rate bonds, put options, zero-coupon bonds, etc.). Chapter 4 elaborates on capital recovery charges, an increasingly popular source of revenue related to capital facilities. Specifically, Chapter 4 discusses the background of capital recovery charges, objectives of an effective charge structure, the advantages and disadvantages of capital recovery charges, methods of calculating and assessing these charges, and other considerations related to their development and implementation. In Chapter 5, the concept of privatization as a method of financing capital facilities is discussed. Specifically, the chapter provides an overview of privatization, the structure of a privatization transaction, the advantages and disadvantages of privatization, several detailed approaches to structuring a privatization transaction, determining feasibility of privatization, privatization procurement and implementation, and state considerations related to privatization. Chapter 6 recommends a process for selecting an appropriate financial plan, considering such factors as short-term and long-term interest rates, costs and time frames of long-term issuance, risks associated with changing market conditions, the degree of public acceptance of the financing program, potential tax law changes, debt capacity, and timing of capital improvements.

Part II of the book deals with water and wastewater pricing. In Chapter 7, an overview of the three-step cost determination and rate setting process is presented, with subsequent chapters dealing with each step of the process. Establishing water and wastewater revenue requirements is discussed in Chapter 8. Specifically presented are the development of revenue requirements under the utility and cash needs approaches, the determination of operating and maintenance costs, and the identification of capital costs. Chapter 9 deals with allocating revenue requirements to various cost components and classes of water and wastewater customers. In Chapter 10, rate design is discussed. Specifically discussed are how to structure a minimum (fixed) and volume (usage) charge, and different types of rate structures that are used by various utilities. Chapter 11 discusses rate-setting trends across the country and presents the results of the 1988 Water and Wastewater Rate Survey, published by Arthur Young & Company.

Numerous complex issues are involved in water and wastewater financing and pricing. For this reason, this book deals primarily with major concepts and approaches to financing and pricing. Several important pricing concepts will be mentioned but not discussed in detail: pricing public and private fire protection; the development of high-strength wastewater surcharges; how to recover the costs of non-revenue water (water theft, billing inaccuracies, water leaks, etc.); how to recover the costs of wastewater infiltration and inflow; the appropriateness of providing discounts or subsidies to certain classes of utility customers (elderly, economically disadvantaged, etc.); and justifying outside-city rate differentials. These topics can be the subject of separate texts and were excluded not because they lack importance, but to focus this book on major components of the financing and pricing process.

Part I
FINANCING
WATER and
WASTEWATER
SERVICES

2

CAPITAL AND FINANCIAL PLANNING FOR WATER AND WASTEWATER UTILITIES

2   CAPITAL AND FINANCIAL PLANNING FOR WATER AND WASTEWATER UTILITIES

Capital facilities represent a major investment by water and wastewater utilities. Supply, treatment, transmission, and distribution facilities are needed to provide potable water to homeowners, businesses, institutions, and industrial customers. Investments in collection, transmission, treatment, and disposal facilities are required for wastewater service. Capital investments are necessary to maintain high-quality service to existing customers and to provide facilities for growth and economic development.

A major challenge to most utilities is to develop an effective long-range capital and financing plan. The plan first of all identifies the type of facilities that are required over a long-range planning horizon for (1) expanding service, (2) upgrading water and wastewater treatment quality, (3) replacing dilapidated and deteriorated water and wastewater infrastructure, and (4) providing for smaller recurring capital needs. In addition, financial requirements related to the capital plan are identified by year, and appropriate sources to finance these capital items are developed. During this development of appropriate sources of financing, economic impacts on utility customers are carefully evaluated.

The capital and financial planning process should be comprehensive to identify the most appropriate capital items in which to invest, phase the purchase or construction of the items appropriately, and ensure that the utility maximizes available financial resources. This chapter discusses the capital and financial planning process and the key issues that a utility should consider during this process. Subsequent chapters in Part I discuss various capital financing alternatives and how the most appropriate capital and financing method is selected.

A. Description of Capital Items

Before discussing the capital and financial planning process, different types of capital items that would be purchased by a utility should be discussed. Figure 2.1 depicts examples of the types of capital items found in a typical water and wastewater utility. Water and wastewater capital items in general are discussed in the following sections.

1. MAJOR TRUNK FACILITIES

These facilities are typically the larger capital facilities that must be constructed in order to provide the "trunk" or "backbone" water and wastewater system. Since these facilities tend to benefit all customers of the system, their related costs should be recovered from all users of the system. For wastewater systems, these facilities would include treatment plants, large mains, outfall lines, major pumping stations, and sludge disposal facilities. For water systems, these facilities would include surface water reservoirs, conveyance systems, wells, water treatment facilities, large mains, and elevated and ground storage systems. These facilities should be differentiated from facilities that benefit a small sector of the service area or small group of customers.

2. WATER AND WASTEWATER EXTENSIONS

These facilities are typically sewer collection mains or water distribution mains that are extended from the trunk water and wastewater system to provide service to a specific part of the service area. They can include mains that are extended to real estate subdivisions or to a specific utility customer, as well as any local service mains within a subdivision. These facilities normally benefit a specific group of customers with related capital costs being recovered from the benefiting customer. It is difficult to generalize that a certain diameter pipe should be classified as an extension. In some jurisdictions, a 6- to 12-inch line could be an extension. In other jurisdictions this size line could be considered a trunk main.

3. WATER AND WASTEWATER SERVICE LATERALS

The cost associated with physically connecting a customer to the system is typically capitalized. The connection is performed by the utility or contracted out by the utility. From local service or subdivision mains, it is necessary to construct laterals to the customer's property line for connecting a customer to the water distribution and sewer collection system. If performed by the utility, labor and material costs in the operating budget are identified and capitalized. If the work is contracted out, the amounts paid to the outside contractor would be capitalized. These capital costs are usually recovered directly from the customer that the service installation benefits.

4. CAPITAL EQUIPMENT AND OTHER MINOR CAPITAL ITEMS

These capital items include equipment components of major capital facilities (motors, pumps, instrumentation, etc.), vehicles (cars, trucks, backhoes, lawn mowers, etc.), furniture, major tools, and stand-alone equipment. The utility must determine the cutoff regarding which minor capital items should be expensed through the operating budget and which of these items should be capitalized for depreciation purposes. For example, most utilities would expense minor laboratory equipment such as test tubes, lab handling equipment, and other similar items. The same utility would likely capitalize major laboratory equipment such as gas chromatographs, sampling equipment, etc. Most utilities have guidelines as to what will be expensed and what will be capitalized. The dollar threshold for capitalizing items usually ranges somewhere between $100 and $500.

5. CAPITALIZED OPERATING COSTS

An example of capitalized operating costs would be staff engineer time for designing a water transmission line. As the engineers charge time to the design project, an appropriate amount of their salary would be distributed through the accounting system to the capitalized transmission line. Material costs, consulting fees, and other operating and management (O&M) costs are occasionally identified with a fixed asset and are capitalized. Once costs are capitalized, the capitalized asset would normally fall in one of the capital categories discussed above.

Each capital item plays a unique role in providing water and sewer service. In addition, the costs of each capital item may be recovered differently. Therefore, capital facilities should be categorized appropriately for accounting, management control, insurance, reporting, and rate-setting purposes. The utility's fixed asset accounting system plays an important role in identifying, updating, and presenting fixed

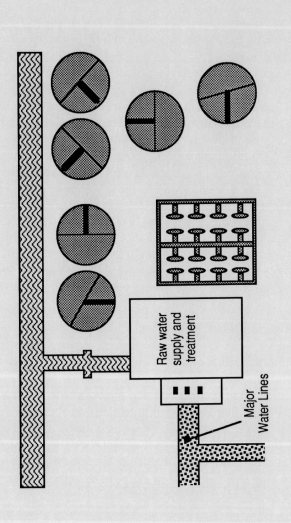

FIG. 2.1A.   WATER CAPITAL COMPONENTS

Raw water supply and treatment

Major Water Lines

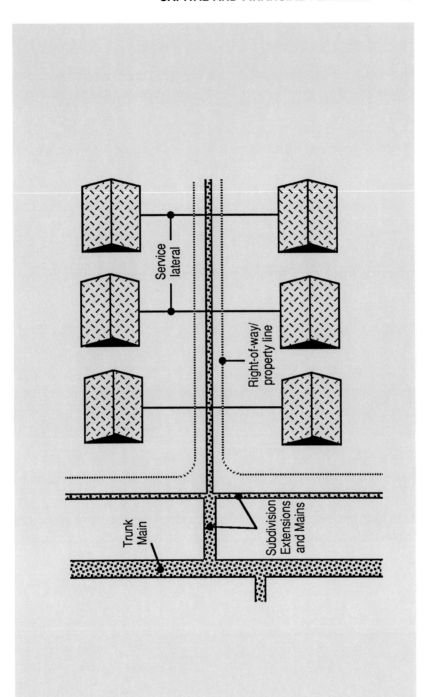

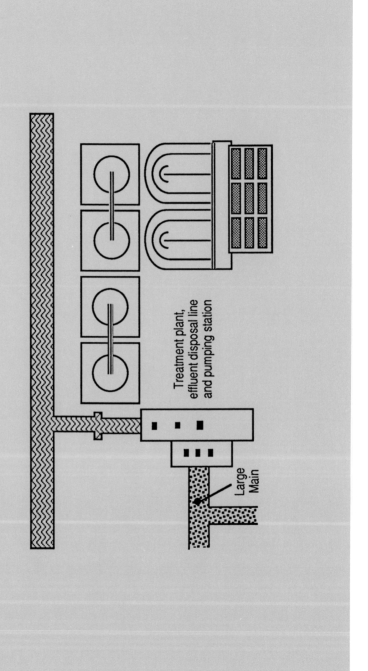

FIG. 2.1B. WASTEWATER CAPITAL COMPONENTS

Treatment plant, effluent disposal line and pumping station

Large Main

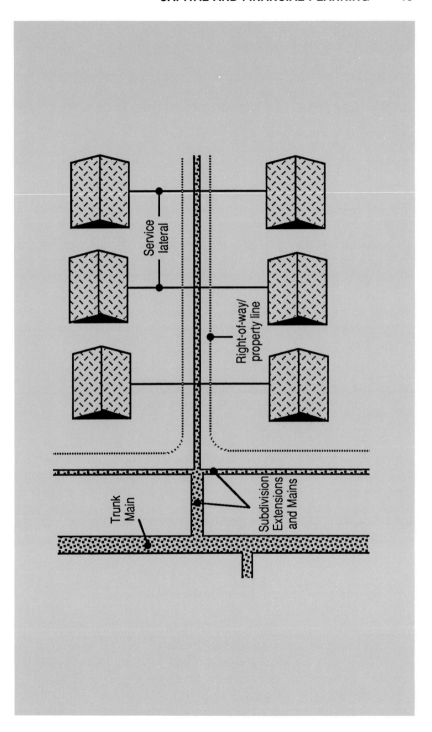

asset cost information for utility managers, governing body members, regulators, and third parties.

B. The Capital and Financial Planning Process

A utility should follow a comprehensive and effective capital and financial planning process, such as the one depicted in Figure 2.2. This process insures that all relevant factors are considered and that the capital plan is consistent with the planning objectives of the utility and the community it serves. The planning process will vary between small utilities and large utilities, between investor-owned utilities and governmental utilities, and between utilities with "steady state" operations (low or stagnant growth) and those addressing rapid growth and economic development. The utility should identify the appropriate participants in the capital and financial planning process and insure that all political, economic, financial, legal, regulatory, and operational issues are addressed. This process is discussed in the remainder of the chapter.

STEP 1: EVALUATE ECONOMIC FACTORS AFFECTING CAPITAL AND FINANCIAL PLANNING

The first step in the capital planning process is to carefully identify and consider those factors which affect capital and financial planning for the utility. Factors should be identified that affect the type and sizing of facilities to be included in the capital plan, the manner in which facilities are financed, and the way costs are recovered from utility beneficiaries and the public. As presented in Figure 2.2, several factors that might affect the capital and financial planning process are discussed below.

- *Customer Demand and Economic Development*
  Probably the most important factor affecting utility capital expansion is growth in demand. As areas are developed or annexed, additional pressure is placed on a utility to provide water and wastewater services. This growth can be caused by economic development (whereby new customers are attracted to a service area), or through specific annexations of additional customers. In addi-

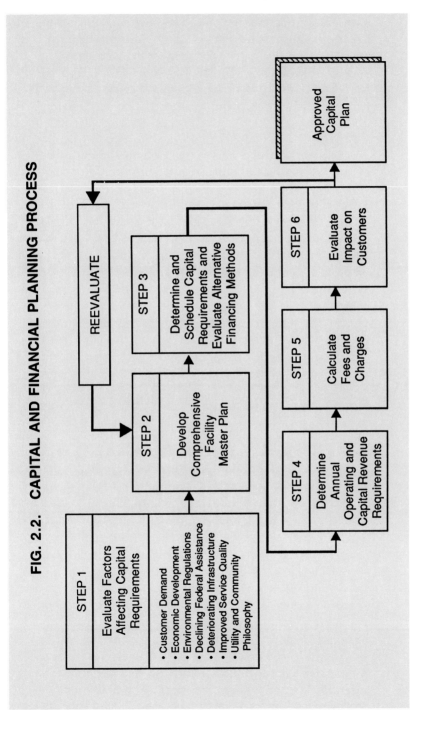

FIG. 2.2.   CAPITAL AND FINANCIAL PLANNING PROCESS

tion, other factors such as per capita water usage, unaccounted-for water (water utility), infiltration/inflow (wastewater utility), price elasticity of demand (how customers will react to water and wastewater rate increases), and long-term environmental impacts must be considered in estimating demand for a particular service area.

- *Deteriorating Infrastructure*
  In many service areas, the water and wastewater infrastructure is old and deteriorating, requiring replacement. Varying conditions from jurisdiction to jurisdiction, such as growth, weather, soil types, elevation, water quality of discharge streams, and water quality of source of supply, require unique infrastructure maintenance. Some areas of the country, such as parts of Arizona, California, and the Sunbelt, are rapidly growing. Utilities serving these areas typically have a large percentage of their capital investment in new facilities. As a result, these communities will not be confronted with major infrastructure deterioration until future years. However, in parts of New England and industrial areas of the Northeast and Midwest it is not uncommon to have facilities built in the late 1800s and early 1900s, requiring major rehabilitation. Replacing specific infrastructure items can be as costly as constructing facilities for expansion.

- *Increasingly Stringent Environmental Regulations*
  Since its inception in the early 1960s, the EPA has been focused on protecting and improving environmental quality. To achieve its goal of upgrading water quality, the EPA developed environmental regulations. These regulations have been imposed upon government- and investor-owned utilities, industries, owners of environmental facilities, and other entities involved in activities that could affect water quality. In addition, state environmental agencies have been formed to enforce federal regulations and in some cases have developed more stringent state requirements. Increased environmental regulations have forced many water and wastewater utilities to upgrade existing facilities or to build new facilities that meet much more stringent environmental limitations.

- *Customer Demands for Improved Water and Wastewater Service*

  At a time when capital facilities have become more costly, customers are demanding improved services from their water and wastewater utilities. Water taste, odor, and appearance are a continual concern to many residential and commercial customers. Industrial customers continue to insist on improved water and wastewater services for their manufacturing processes in order to improve the quality of their industrial output. Wastewater odors and the health impact of insufficiently treated wastewater discharged into streams and waterways are major quality issues. As a result, capital facilities are often included in a capital program for addressing customers' quality concerns.

- *Declining Federal Assistance*

  At a time of more stringent environmental regulation, the EPA and the states are moving toward having local governments finance water and wastewater facilities without federal and state grant assistance. Public Law 92-500 (the Water Pollution Control Amendments of 1972) was the largest public works grant program in our country's history. This program has assisted wastewater owners (called "201 grantees") in building facilities to address more stringent environmental quality requirements. In the 1980s, funding for this public works program has been significantly reduced, but the environmental regulations which 201 grantees must address are still very much a reality. In many cases, certain wastewater utilities must address environmental regulations without federal and state grant assistance. To provide some financial relief for the diminishing grant program, EPA is providing funds to states for establishing state revolving loans programs. The primary objective of these programs is to provide a source of low-interest loans to 201 grantees for financing their wastewater construction requirements. Even though revolving loan programs provide some assistance to government utilities that qualify, utilities must repay these loans as if a bond issue were used. As a result, utilities are more selective in identifying which facilities they are willing to finance. In addition, in an attempt to reduce the economic impact of declining federal and state assistance, alternative funding methods, consolidation, and the construction of regional facilities will be encouraged in the future.

- *Utility and Community Philosophical and Legal Restrictions*
  A utility's philosophy and its community's planning goals play an important part in identifying what facilities should be included in the capital plan, how these facilities should be financed, and how related costs should be recovered from the utility's customers. The utility must address several important philosophical and legal questions:
  –Should growth be required to partially or totally pay for itself?
  –Are certain financing techniques disallowed by state law (adjustable rate bonds, commercial paper, lease/purchase, tax increment financing, etc.)?
  –Should facilities be oversized to accommodate growth and how should this oversizing be financed?
  –Should existing customers be required to share in the cost of system expansion?
  –Should certain classes of customers be subsidized?
  –Is conservation an important issue in the service area and should financing mechanisms be developed to encourage conservation?
  –Should the utility channel growth to a particular corridor of the service area?
  –Should less risky conventional or more risky innovative financing techniques be adopted by the utility?
  –Does the community want to be competitive in attracting economic development to its area?

These are a few of the questions that must be addressed in determining the most appropriate way of structuring a capital and financial plan.

STEP 2: DEVELOP A COMPREHENSIVE FACILITY MASTER PLAN

In order to develop a comprehensive water and wastewater facility master plan, all the items noted in Step 1 must be considered. The master plan identifies the capital facilities required for expansion, upgrade, and rehabilitation of the water and sewer system. Since construction time frames can be lengthy and economies of scale can result from use of larger facilities to meet long-term demand, the planning horizon of a master plan is usually 20 to 30 years. In some cases, the planning time can extend to 50 years or longer. The master plan must be comprehensive and is essential for securing appropriate financing.

During the master planning process, an engineer (either a staff or a consulting engineer) evaluates alternative technological solutions and selects an appropriate configuration to address a community's requirements. In some master plans, detailed support for the proposed capital program can be extensive. The plan might also include comprehensive demand studies as well as a thorough evaluation of the environmental, financial, legal, political, and operational impacts of adopting the proposed master plan. The facility master plan will normally provide broad estimates of cost for proposed capital facilities, with cost estimates for immediate facilities more precise than those for long-range facilities. In many jurisdictions, the master plan is divided into five-year increments over the long-term planning horizon.

STEP 3: DETERMINE AND SCHEDULE CAPITAL REQUIREMENTS AND EVALUATE ALTERNATIVE FINANCING METHODS

After a facility master plan is developed for a long-range planning horizon, the next step is to identify the first increment of capital needs that will have to be addressed. These needs are included in a comprehensive capital budget, usually developed for a five-year period. In addition, the five-year budget might include the costs of water and wastewater extensions and the costs of minor capital items not identified in the master plan. The year and time when capital facilities become operational, as well as construction time frames associated with capital facilities, must be carefully considered. Recommended financing methods are included as part of the capital improvement plan.

In financing the capital program, several financing sources could be considered:

- short-term financing (loans, anticipation notes, and commercial paper)

- grants
- bonds
- capital recovery charges (impact fees)
- developer contributions
- assessments
- privatization
- lease/purchase
- dedicated capital funds
- operating revenues
- interest revenue

Short-term financing and bonds, capital recovery charges, and privatization are discussed in detail in Chapters 3, 4, and 5, respectively.

Table 2.1 presents an example of a five-year capital schedule for a wastewater capital improvement plan. In the example, a major wastewater treatment plant expansion and upgrade are planned for the five-year planning period. In addition, a pumping station upgrade and expansion, sewer line extensions and replacements, and a composting facility are included in the plan. "Pay as you go" capital items (vehicles, furniture, and equipment) are also included in the plan, even though in some utilities these are funded through the operating budget.

In Table 2.1, a combination of traditional short-term and long-term financing is employed. In addition, grant funding and privatization are used as sources to address certain capital needs. Developer contributions are used to finance specific facilities from which the developer would benefit. In a subdivision where residents are on septic tanks, extension expansion and local service line construction are financed through assessments. Interest income on long-term investments also provides a source of funds for the capital programs. In some utilities, current revenues (user charges, etc.) are used to recover a substantial portion of capital requirements. In most utility financing plans, current revenues are a primary source of recovering capital costs when other financing sources have been exhausted or have been deemed inappropriate.

Alternative financing scenarios may have significantly different annual revenue requirements associated with them. For example, using short-term financing might delay the recovery of capital costs to later years. However, in later years, annual revenue requirements could be significantly higher. If capital replacement or capital expansion have been sufficiently funded, the short-term and long-term debt market could be avoided altogether. In today's complex financial market, it is likely that a utility will utilize a variety of traditional and innovative funding services to maximize the financing objectives of the utility.

The full capital improvement plan would include additional detail (locations, maps, project descriptions, demand projection, etc.) for supporting the facilities included. In addition, it is typical to have separate plans for water and wastewater, with joint facilities (administration buildings, joint equipment and vehicles, etc.) integrated into the plan.

Before a financing plan can be finalized, the utility must understand the economic implications of the financing plan.

The remaining steps of the capital and financial planning process will assist the utility in evaluating economic impacts and finalizing the capital improvement plan.

STEP 4: DETERMINE ANNUAL OPERATING AND CAPITAL REVENUE REQUIREMENTS

The next step in the capital and financial planning process is to translate a proposed capital improvement plan into annual revenue requirements for the five-year period. Revenue requirements should include costs associated with relevant financing techniques identified in Table 2.1: (1) proposed bonds, (2) existing financing requirements (existing bonds), and (3) operations and maintenance costs associated with existing and proposed facilities.

Increased operating costs associated with new facilities as well as proposed operating costs of existing facilities need to be identified. In some cases, certain operating costs can be reduced by building new facilities, e.g., a more technologically advanced wastewater treatment plant operating with less staff and lower energy costs than the existing plant.

Table 2.2 identifies annual revenue requirements associated with the proposed capital plan:

- operating and maintenance costs on existing and proposed facilities
- existing and proposed debt service
- interest costs on bond anticipation notes
- contributions to capital replacement fund
- contributions to capital expansion fund

Table 2.1 Capital Planning Process: Identification of Wastewater Capital Requirements and Sources, FY1989 Through FY1993 (in $1000's)

Projected Demand (in 1000 gallons at plant)	4,200,000	4,600,000	5,100,000	5,700,000	6,300,000	15,900,000
Capital Requirements	FY1989	FY1990	FY1991	FY1992	FY1993	TOTAL
Expansion of Whites Bridge Road Wastewater Treatment Plant	$3,500	$2,200	$ 500	$	$	$ 6,200
Upgrade of Andrews Highway Wastewater Treatment Plant		1,500	2,250			3,750
Construction of the Gaves Pumping Station	550					550
Upgrade and Expansion to Friendfield Pumping Station				200		200
Replacement of Wedgefield Sewer Line					600	600
Belle Isle Garden Sewer Extensions		450				450
Lateral Construction for Kensington Subdivision			300	500		800
Construction of Composting Facility				4,000	3,000	7,000
Minor Capital Items		600	600	600	600	2,400
• Backhoe	150					150
• 3 5-Ton Trucks	210					210
• 1 1/2-Ton Truck	10					10
• 1 Vacuum Truck	75					75
• Miscellaneous Furniture	18					18
• Miscellaneous Equipment	75					75
TOTAL CAPITAL NEEDS	$4,588	$4,750	$3,650	$5,300	$4,200	$22,488

Table 2.1 Continued

Capital Sources						
Grants		$1,000	$1,250			$ 2,250
Short-Term Borrowing (BANs)		500	1,000			1,500
Capital Recovery Charges	538	600	600			1,738
Long-Term Bond Proceeds	$3,500	2,200	500			6,200
Developer Contributions		450				450
Capital Expansion Fund	550					550
Capital Replacement Fund				$ 500	$ 900	$ 1,400
Privatization				4,000	3,000	7,000
Interest Revenue				300	300	600
Assessments			300	500		800
TOTAL	$4,588	$4,750	$3,650	$5,300	$4,200	$22,488

Table 2.2   Capital Planning Process: Determination of Revenue Requirements for User Charges, FY1989 Through FY1993 (in $1000's)

	FY1988	FY1989	FY1990	FY1991	FY1992	FY1993
Revenue Requirements	(Current)					
O&M Costs						
Existing Facilities	$1,740	$1,800	$1,870	$1,940	$2,020	$2,120
Proposed Facilities				450	550	950
TOTAL O&M REQUIREMENTS	$1,740	$1,800	$1,870	$2,390	$2,570	$3,070
Capital Costs						
Debt Service (Including Coverage)						
• Existing	2,000	2,000	2,000	1,950	1,950	1,900
• Proposed		310	620	620	620	620
Interest on Bond Anticipation Notes			25	125		
Privatization Service Contract						1,050
Contributions to Capital Expansion Fund	475	500	525	530	540	555
Contributions to Capital Replacement Fund	235	250	280	290	320	350
TOTAL CAPITAL REQUIREMENTS	$2,710	$3,060	$3,450	$3,515	$3,430	$4,475
TOTAL REVENUE REQUIREMENTS	$4,450	$4,860	$5,320	$5,905	$6,000	$7,545
% Increase (Decrease) in Annual Revenue Requirements		9.2%	9.5%	11.0%	1.6%	25.8%

The annual percentage increase or decrease of revenue requirements to be recovered through user charges is also presented in Table 2.2. These increases or decreases will be useful in determining whether capital programs should be scheduled differently or whether rate increases should be phased in over time. These percentage changes, however, are not projected user charge adjustments. To determine user charge adjustments, the impacts of increased or decreased user demand must be considered.

STEP 5: CALCULATE FEES AND CHARGES

After annual operating and capital revenue requirements are determined, customer fees and charges can be determined. Capital recovery charges such as impact fees and front-foot assessments would be based on the capital program or another appropriate method for valuing certain facilities. Monthly rates and miscellaneous user fees would normally be based on annual revenue requirements after deducting revenue offsets. Chapters 7 through 10 discuss the development of these charges in detail.

At this point, fees and charges are typically estimated to determine the preliminary feasibility of the capital plan. If economic impacts are too severe, then modifications to the plan may be required. The remaining steps in the financial planning process are focused on finalizing the capital plan.

STEP 6: EVALUATE IMPACT ON CUSTOMERS

The next step in the capital and financial planning process is to evaluate the impact that a proposed capital plan will have on customers and others required to pay under the various financing scenarios. If general obligation bonds are used, with the debt service on the bonds being paid through the general fund of a community, then the impact on property taxes is carefully evaluated. If user charges are financing the majority of the capital plan, then the impact on the water and sewer customer rates is carefully considered. If capital recovery charges are considered for financing a significant portion of customer growth, then the impact of attracting or distracting customers to a particular area is evaluated.

Table 2.2 identified revenue requirements to be recovered through user charges. The impact of using user charges to recover costs of part of the capital program is presented in Table 2.3. It is important to choose customer classes, at various consumption levels, that will pro-

Table 2.3   Capital Planning Process: Wastewater Customer Monthly Impact Analysis, FY1990

Customer Type	Current Charges	Proposed Charges	Increased (Decreased)	% Increase (Decreased)
Residential – Inside City (⁵/₈″ meter)				
0 gallons	$ 8.00	$ 8.00	$ 0	0%
4,000 gallons	12.00	13.00	1.00	8.3
8,000 gallons	16.00	18.00	2.00	12.5
12,000 gallons	20.00	23.00	3.00	15.0
16,000 gallons	24.00	28.00	4.00	16.7
20,000 gallons	28.00	33.00	5.00	17.9
Residential – Outside City (⁵/₈″ meter)				
0 gallons	12.00	12.00	0	0
8,000 gallons	24.00	27.00	3.00	12.5
Commercial				
Laundry (2″ meter – 500,000 gallons)	620.00	745.00	125.00	20.2
Service Station (1¹/₂″ meter – 350,000 gallons)	400.00	487.00	87.50	21.9
Restaurant (1¹/₂″ meter – 400,00 gallons)	450.00	550.00	100.00	22.2
Beauty salon (1″ meter – 100,000 gallons)	120.00	145.00	25.00	20.8
Industry				
A – 15,000,000 gallons	15,000.00	18,750.00	3,750.00	25.0
B – 5,000,000 gallons	5,000.00	6,250.00	1,250.00	25.0
C – 2,000,000 gallons	2,000.00	2,500.00	500.00	25.0
Municipal Customer – 30,000,000 gallons	30,000.00	37,500.00	7,500.00	25.0

vide valuable insights into how the majority of customers will be affected by various capital programs. In addition, it may be beneficial to evaluate rate impacts during different times of the year to provide a realistic picture of how different customers will be affected. It is also important to show rate impacts for several years to evaluate the long-term impacts of various capital plans.

In some cases, the preliminary capital plan produces such a significant impact on customers that it will have to be reevaluated. Certain capital projects may be abandoned altogether; certain projects might have to be restaged; certain projects might have to be downsized or reconfigured. In making these trade-offs, however, it is important to consider economic, legal, operational, regulatory, and political factors. As a result of going through this process several times, the utility staff and governing body can develop a capital plan which optimizes community objectives given the constraints imposed upon the utility.

In developing a comprehensive financial plan, significant amounts of data have to be processed. In addition, there are numerous financing scenarios that could be considered in addressing the capital needs of a particular community. For these reasons, it is advantageous to use the microcomputer for evaluating alternative financial planning scenarios. Once the capital plan is approved, the microcomputer can be used in evaluating how effective the capital plan has been by comparing actual versus budget amounts. The microcomputer also allows the utility to employ flexible budgeting, as actual results will most likely differ from budgeted amounts. Placing the capital planning model on the microcomputer will also be useful in performing future capital planning updates as better information regarding future projects becomes known.

The remainder of Part I of this book discusses common financing methods that could be considered in financing capital facilities. Particular attention will be given to:

- short-term financing
- bonds
- capital recovery charges
- privatization

3 BONDS, SHORT-TERM FINANCING, AND CREDIT ENHANCEMENTS

3   BONDS, SHORT-TERM FINANCING, AND CREDIT ENHANCEMENTS

In Chapter 2, bonds and short-term financing were identified as revenue sources for financing capital requirements. Utility management and governing boards are confronted with the complex task of evaluating alternative financial plans with long-range and short-term debt instruments. At one time the process was relatively simple — estimate capital requirements and either issue traditional tax-free (1) revenue or general obligation bonds, or (2) short-term notes, until permanent financing became attractive. In today's complex financial environment, however, it is advantageous to evaluate alternative financing instruments.

A. History of Changes in Financing Methods

Over the past few years, utilities have seen significant change in the tax-free debt market. Prior to the early 1980s, government utilities could issue debt at relatively low interest rates. Innovative financing methods were virtually nonexistent, with most utilities using traditional methods for financing (general obligation and revenue bonds). As the rate of inflation increased, tax-free interest rates also increased. At that time, many governmental utilities were reluctant to enter the market. In response to increasing interest rates, investment banking firms introduced innovative financing techniques to lower interest rates. To receive lower interest rates than under traditional financing methods, however, bond holders had to be given certain "sweeteners," such as interest rate adjustment periods, redemption and conversion features, credit facilities (letters of credit and bond insurance), interest rate payment exchange agreements ("swaps"), and other innovative financing techniques. These features protected the investor if major changes in the market occurred.

As a result of economic recovery in the mid-1980s, investor confidence was restored in long-term and short-term investment instruments. Interest rates approached pre-1980 levels for long-term and short-term financing. The mid-1980s became a period of significant water and wastewater expansion, with most of this expansion financed by tax-exempt debt. It also became a time when most debt issued in the late seventies and early eighties was refinanced by new debt. Refinancing took place because interest rates decreased significantly. In most cases, the lower debt service payments far outweighed the issuance cost of refinancing the existing debt.

The Tax Reform Act of 1986 and the Deficit Reduction Act of 1987 were passed, attempting to generate additional federal tax revenues to reduce the federal deficit. A major part of the legislation dealt with tax law changes related to taxable and tax-exempt bonds, and how the proceeds could be used. The definition of "public purpose" to gain tax-exempt status for infrastructure bonding became more restrictive. At the same time, the law required many bondable projects to be classified as "private activity," which severely restricted the tax-exempt status of the bonds. Over the past couple of years, government utilities have evaluated their borrowing philosophy in a changing tax environment. There has been a concern that gaining tax-free status for debt is more difficult. As a result, issuers have been uncertain as to how attractive their debt might be to potential investors. Immediately after the passage of the 1986 and 1987 legislation, utilities were cautious to proceed into short-term and long-term financing while the tax-free status of their debt was still being determined by bond counsels. At the time this book was published, investor confidence has been largely restored, with government utilities moving more aggressively to the bond market.

If a utility decides to issue debt, its first decision is whether to go to the short-term or long-term debt market. In the current market, short-term debt instruments that are available to the utility would include:

- traditional short-term notes
  - bond anticipation notes (BANs)
  - grant anticipation notes (GANs)
  - tax anticipation notes (TANs)
  - revenue anticipation notes (RANs)
- tax-exempt commercial paper (TECP)

- line of credit

If the utility decides to go immediately to the long-term debt market, it is faced with the decision between traditional and innovative financing vehicles:

- traditional long-term financing options
  - –general obligation bonds
  - –revenue bonds
  - –double-barrel bonds
- innovative long-term options
  - –daily adjustable tax-exempt securities (DATES)
  - –variable rate demand bonds or obligations (VRDBs or VRDOs)
  - –adjustable rate bonds (ARBs)
  - –zero coupon bonds
  - –interest rate payment exchange agreements (swaps)
  - –certificates of participation
  - –leasing

This chapter describes the major short-term and long-term financing instruments and discusses the advantages and drawbacks of each method. In addition, the chapter discusses how short-term borrowing can be more attractive to utilities through the use of credit enhancements.

B. Short-Term Financing

A traditional objective of utility debt financing has been to match the term of the financing approximately with the useful life of the asset being financed. As a result, the cost of a major facility asset can be recovered from its users over the useful life of the facility. For example, a water treatment plant with a useful life of 25 to 30 years before major rehabilitation may be financed by a 30-year bond.

In the early 1980s, long-term debt financing was made more difficult and costly by record high interest rates and volatile fluctuations in these rates. In response to the need for increased debt financing at a time of unstable markets, a variety of short-term debt techniques was developed to achieve lower rates. These techniques enable issuers to issue debt with lower interest rates by using a short-term "put" feature to make the securities more attractive to investors. Some of these instruments were virtually unknown to local and state governments until several years ago. In addition, financing provided the utility flexibility to convert to a long-term instrument when the market became more favorable for long-term debt. Basically, several of these instruments have been adaptations from the private sector, which has traditionally used a variety of short-term capital financing techniques.

During the '80s, innovative and traditional short-term financing vehicles have been very popular. Discussed below are examples of these short-term techniques, issued in anticipation of some permanent source of long-term financing. These techniques include tax-exempt commercial paper; variable rate demand notes; and traditional fixed rate demand notes.

1. TRADITIONAL FIXED RATE NOTES

Fixed rate notes are short-term notes usually issued for a period of one to three years. They have been used traditionally as short-term financing for major facilities and are typically issued in denominations of $5,000. Since the interest rate is fixed for the amortization period, there is not a provision for the notes to be called by the purchaser of the certificate, and there are no optional redemption features. Credit support from a commercial bank may be required to secure the payment of the notes at the time of redemption. Credit support is normally required because there is likely no revenue source at the time the short-term funds will be used. Notes are issued "in anticipation of" the revenue source materializing. Because of the fixed nature of the short-term vehicle, the interest rate on traditional fixed rate demand notes is higher than that of both variable rate notes and commercial paper.

Common fixed rate notes would include:

- bond anticipation notes (BANs) — issued in anticipation of the sale of long-term bonds

- grant anticipation notes (GANs) — issued in anticipation of the receipt of state or federal grant funds; GANs have been a typical short-term debt instrument used by grantees under the 201 wastewater construction grants program (provided for by the Water Pollution Control Amendments of 1972).

- tax anticipation notes (TANs) — issued in anticipation of the receipt of taxes to be collected from taxpayers within a certain governmental taxing district. TANs would be issued, for example, to finance a local service collection system for a particular community in anticipation of receiving tax funds from residents benefiting from these facilities.

- revenue anticipation notes (RANs) — issued in anticipation of revenues to be generated by an issuing utility. RANs could be used in a startup utility where revenues are virtually guaranteed once facilities are constructed and service initiated.

2. TAX-EXEMPT COMMERCIAL PAPER

Tax-exempt commercial paper is defined as short-term, unsecured promissory notes, backed, for liquidity purposes, by a line of credit from one or more banks. In some cases, a letter of credit is received for TECP. Maturities normally range from one to 270 days, with an average maturity of approximately 30 to 45 days. Interest rates for tax-exempt commercial paper can be as low as 50% of bank prime interest rates. As a result, this becomes an attractive short-term financing vehicle for local governments when short-term note financing is in the neighborhood of 60% to 70% of the prime rate. TECP is often the least costly method of interim financing available to government utilities. Even with 50 to 75 basis points added for necessary commercial bank liquidity and/or credit support, TECP can offer the lowest effective cost possible to eligible borrowers. TECP also offers the shortest fixed maturities in the tax-exempt market and, therefore, the greatest flexibility. As a result, TECP can be sold on a specific date that funds are required, with maturity dates established to coincide with dates when revenues such as taxes, bonds, grants, or user charges are anticipated. The short-term fund issues can also be sold in smaller amounts that would be impractical in traditional bond markets.

Another advantage of TECP is that it can be used to address seasonal peaks when working capital tends to be at its lowest. A further benefit is that the proceeds from the sale can be received on the day of the sale. TECP is contrasted with long-term debt issues in which funds may not be received until well after the date of sale.

For all of its advantages, a certain risk is associated with TECP. As with any short-term financing technique, there is always a chance that by the time the project is ready for long-term financing, long-term interest rates may rise. The risk is mitigated, however, by the savings that would be achieved through TECP at the time that long-term interest rates are increasing. These savings would serve to increase the rates to which the long-term bond take-out would have to rise before an issuer would be worse off by having implemented a TECP program.

Another disadvantage of TECP is that the issuer could become too dependent upon "rolling over" TECP, thus leading to financial instability. Finally, because of the liquidity of TECP, the market demands the highest short-term credit ratings for TECP issuers. If the issuers are not sufficiently strong, some form of credit facility (discussed below) from a commercial bank will likely be required. Some states, however, do not allow the issuance of TECP. In other states, the issuance of TECP is severely restricted. As the TECP market is essentially continuous, replacing maturing debt through debt issuance may not be permitted under existing state regulations governing municipal debt management. Depending on the particular state regulation, TECP may not be available to the municipalities or may have to be paid off from current revenues at least once annually.

3. TAX-EXEMPT VARIABLE RATE DEMAND NOTES

Tax-exempt variable rate demand notes (VRDNs) are sold to investors who, after some period of time, have the right to "demand" payment of the face value of the notes. While the investors hold the notes, their interest rate is tied to market conditions. Because of the high liquidity of variable-rate demand notes, they are very attractive to investors and tend to carry a lower rate of return than less flexible, more traditional financing instruments.

In order to insure that the issuing government utility can meet demand payment criteria, the utility would normally enter into a credit facility with a commercial bank. If the variable rate demand note is called, the issuer has the flexibility of paying off the note with funds at hand or using funds from the credit facility.

Table 3.1 summarizes the advantages and disadvantages of various short-term financing techniques.

C. Credit Facilities for Short-Term Financing

Most issuers require that a commercial bank provide a credit facility to support variable rate demand notes, tax-exempt commercial paper, and on occasion, fixed rate demand notes. A credit facility is

Table 3.1   Short-Term Financing Instruments

Type Instrument	Advantages	Disadvantages
Fixed Rate Notes	• No risks exist for potentially higher rates. • Less security is required than with other short-term instruments. • Issuance period can be up to five years. • Notes can be issued in lower denominations ($5,000).	• Interest rates are typically higher than for other short-term financing techniques. • Specific maturity dates may not coincide directly with a long-term financing schedule.
Tax-Exempt Commercial Paper (TECP)	• TECP usually has the lowest available interest rates in the debt market. • TECP can be issued with specified maturity date to coincide with long-term financing schedule. • There is no "put" requirement, interest rate adjustment period, conversion features, or redemption features. • Only the amount needed is borrowed initially.	• Letter of credit is typically required. • Issuer can be prone to "roll over" debt, creating some instability, to the financing worthiness of the issuer. • There is no ability to convert to a fixed mode. • Maturing time frame is short (1 to 270 days). • Administrative time and overview can be more extensive than other short-term financing instruments.
Variable Rate Demand Notes	• Interest rate is typically lower than fixed rate notes. • Notes can be issued for periods greater than TECP. • Notes can be converted to a fixed mode. • Minimal, if any, program administration is required.	• Letter of credit is typically required. • Rates are normally somewhat higher than TECP. • Notes have a "put" feature within a specified call period. • Interest rate adjustment period can be frequent.

security to the investor that, if the issuer is unable to pay required interest and/or principal on a note, then the "securing" institution will guarantee the payment of relevant amounts. As a result of guarantees from this credit facility, the issuer can receive favorable interest rates. The credit facility is usually either a letter of credit or a line of credit.

1. LETTER OF CREDIT

For an issue secured with a commercial bank letter of credit, the bank basically guarantees payment of debt service including accrued interest on the securities. The letter of credit is issued to the bond trustee and represents an irrevocable obligation on the bank's part to pay the trustee, upon demand, amounts representing principal premium and interest on the securities when due. The letter of credit may also provide a liquidity facility, providing funds at a predetermined rate, if bonds are put and not resold by the remarketing agent. The letter of credit generally allows floating rate securities of different issuers to trade interchangeably. The rating of a letter of credit is considered at the time the issue that is being secured is being rated.

Several factors affect the cost or fees on a letter of credit. Fees are subject to negotiation and are based upon the transaction's size and complexity and the credit of the issuer. Certain factors influence the letter of credit fee: (1) legal fees for bank counsel, issuer's counsel, and bond counsel; (2) compensating balance requirements, covenants, coverage ratios, and reserves; (3) placement fees for the one-time issuance of the letter of credit; (4) fees for maintaining the loan accounts; and (5) annual charges for payment of the trustee and paying agency.

Typical fees range from 0.25% to 1% per annum of the face value of the letter of credit, plus accrued interest. The fee is also based upon the term of the letter of credit. Because the securing bank must honor demands under the letter of credit, the longer the term of the letter of credit, the greater the risk assumed by the bank. It should also be noted that a transaction can be structured so that a letter of credit is a part of short-term financing, with the letter of credit dissolved upon permanent financing.

2. LINE OF CREDIT

A line of credit is simply a promise by a commercial bank to lend funds, subject to certain conditions, at a given rate of interest for a certain period of time. A line of credit does not obligate the lending bank to pay debt service on the bonds, nor does it upgrade the rating of the bond issue with the bank's financial strength. A line of credit is usually established for a one-to three-year period, with the line of credit being terminated upon notice by the lending bank. The line of credit is typically not considered as secure as a letter of credit. As a result, short-term vehicles secured by a letter of credit tend to be floated at lower interest rates than those backed by a line of credit. Fees for a line of credit tend to range from 0.125% to 0.375% per annum of the face value of the security.

The major credit rating agencies, including Standard and Poors Corporation and Moody's Investment Services, usually rate tax-exempt commercial paper and fixed rate notes. Variable rate demand notes typically receive two ratings—one on the likelihood of repayment of principal and interest when due, and a second which addresses the quality of the demand feature of the note. If a letter of credit is provided with the short-term financing vehicle, then the rating agency considers the credit rating of the insuring bank. If a line of credit is used, the rating is then based on the underlying credit of the issuer after taking into account the liquidity support of the line of credit.

D. Conventional Long-Term Financing Methods

Conventional financing with short-term fixed rate demand notes and long-term bonds is not the automatic choice it once was. This combination, however, still remains the prevalent financing method in the tax-exempt market.

General obligation bonds and revenue bonds provide the basis for most conventional financing. Double-barrel bonds are like a general obligation bond but are secured by the revenues of a particular enterprise, such as water or wastewater operations. A moral obligation bond is like a revenue bond with a moral, but not legal, obligation of another governmental entity to secure the bonds in the case the issuer cannot meet debt service requirements. These different debt instruments are discussed below.

1. GENERAL OBLIGATION BONDS

General obligation bonds are secured by the full faith and credit of an issuing institution. This institution typically has taxing powers and has the capability of levying taxes to support payments of debt obligations. General obligation bonds have a major advantage in that they are backed by the taxing ability of the government entity, and this credit is usually the strongest security pledge available to an issuer at the lowest net interest cost. In addition, the issuance of general obligation bonds is usually simpler and frequently less costly than other types of debt.

There are disadvantages to general obligation bonds, however. In order to issue general obligation debt, legislative and/or voter approval is required. This process is likely to be time-consuming, possibly delaying work on a project, with no guarantee of successful approval of the bond. In addition, governmental issuers have practical or legal "debt limits" for the amount of general obligation debt they are able to issue. As a result, financing large utility capital expenditures through general obligation debt severely dilutes the ability of the utility to issue future debt. Extensive use of general obligation debt may also endanger the issuer's credit rating. However, in some states, like Massachusetts, general obligation debt for enterprise activities, like water and wastewater, are exempt from debt ceilings.

2. REVENUE BONDS

The second option of conventional financing would be the use of revenue bonds. With revenue bonds, interest and principal are payable solely from the revenue generated from a specific project or utility. In most states, the bond holders do not have recourse to have taxes levied to pay required debt service. The major advantage of revenue bonds is that they protect the general obligation debt capacity for other projects. Revenue bonds are generally tax-free and would sell at interest rates below taxable debt. In addition, revenue bonds can have greater flexibility in market timing, can reduce risk if they are presold, and can have greater flexibility in timing the repayment of principal. Revenue bonds can also be used in situations where general obligation debt may not be allowed.

When compared with general obligation bonds, there are certain disadvantages to revenue bonds. Issuance costs tend to be higher using revenue bond financing. Management, legal, financial, consulting, and engineering fees, along with other issuance costs, generally increase the issuer's cost. Interest rates tend to be higher for revenue bond issues, primarily because security is weak with the lack of a general obligation pledge; there is a greater degree of credit risk associated with revenue bonds than with general obligation debt. Since debt service on revenue bonds is secured by the revenue stream generated by a particular utility project, reduction or discontinuance of that revenue stream could result in a default on those bonds (provided that there is not a credit facility associated with the bonds). Finally, revenue bond indentures may require that all the project's net revenues (revenues after payment of operating and maintenance costs) first be applied either to reducing the outstanding debt or to creating reserve funds for the same purpose.

In most cases, revenue bond indentures require that several reserve funds be established. These reserve funds provide additional security to the investor that adequate funds will be available for operations, ongoing capital requirements, and debt service to be paid to investors. Typically, reserve funds that would likely be required in a revenue bond issue include:

- debt service amortization fund—a holding fund established to pay annual or semiannual debt service payments. The escrow is funded by monthly contributions equal to some percentage of the debt service payment.

- debt service reserve fund—an amount typically equal to the maximum annual debt service over the amortization period; this amount is normally bonded and serves to secure the debt service payment to bond holders. In some cases, the reserve fund is "built up" by user charge contributions over the first couple of years during the amortization period. Recently, municipal bond insurance companies have offered insurance policies as substitutes for cash reserve funds.

- construction reserve fund—bond proceeds which are segregated for construction purposes. Interest on these bonds typically remains within the construction fund, and along with construction fund proceeds is used to pay construction contractors. With the passage of the Tax Reform Act of 1986, interest income in excess of the yield on bonds must be rebated to the federal government if the issue exceeds $5,000,000 in a calendar year.

- renewal and replacement fund—amounts reserved for replacing certain facilities in case of an emergency or unexpected event. The fund can also be used to finance major maintenance and replacement items that are part of the capital improvement plan. The amounts for this fund are typically established by the consulting engineer.

• operating fund — an amount (usually about one-sixth of annual operating costs) that provides funds for routine working capital purposes. An operating fund would likely be bonded in a startup utility operation where little or no working capital is available during initial operations.

• insurance reserve — an amount that is restricted for purposes of self-insurance or supplementing existing insurance coverage for items not normally covered by traditional insurance policies.

• arbitrage rebate fund — since tax-exempt issuers are subject to repaying arbitrage income under certain circumstances, this fund secures the repayment over the arbitrage period.

Reserve funds are typically funded as part of the bond issue or through coverage over a predetermined period of months.

Revenue bond issues usually have coverage requirements as a part of the bond indenture period. Specifically, the issuer is pledging to establish rates and charges at a level so that after payment of operating expenses and debt service, there is a sufficient "cushion" or reserve to secure the payment of the debt service. Coverage requirements are established at a level usually between 1.1 to 1.3 times greater than debt service. A utility's rates should be structured to recover amounts greater than coverage requirements to insure that default will not occur at times of lower revenue-producing periods.

An industrial revenue bond (IRB), sometimes referred to as an industrial development bond (IDB), is a special type of revenue bond. An IRB is issued by a public entity on behalf of a private sector company desiring to finance certain capital improvements. The IRB is tax-exempt, and provides an economic benefit over traditional taxable debt to a private sector firm. The government agency benefits in that the facility that is to be built usually provides for economic development, employment, or a needed service to the governmental service area. For example, IRBs could be issued to finance a privatized wastewater treatment plant, a manufacturing concern to provide for employment, or a privatized water treatment facility to allow for economic development in a depressed area.

With the passage of the Tax Reform Act of 1986 and the Deficit Reduction Act of 1987, the use of IRBs has become severely restricted. In particular, states are limited as to the amounts of IRBs that can be issued (based upon a cap). In addition, the types of facilities for which IRBs can be used have also become much more limited.

3. DOUBLE-BARREL BONDS

A "double-barrel" bond can be viewed as a hybrid of a revenue bond and a general obligation bond. These bonds are secured in a two-tiered manner: (1) the first source of funds used to meet debt service payments on the bonds is derived from a designated revenue source generated by the project, such as user charges; and (2) if the revenue stream is inadequate to pay debt service, the general tax revenue of the governmental entity is to be used. Essentially, the double-barrel bond is a security that ordinarily relies on a project, but a general obligation pledge secures the payment of debt service. The double-barrel bond carries almost the same credit rating as a general obligation bond and will trade close to if not the same as a general obligation market interest rate. As a result, the major advantages of a double-barrel obligation are similar to those associated with the general obligation bonds.

The key question to be answered in evaluating whether a double-barrel bond should be issued relates to how their use will affect the governmental entity's credit and if voters will support the bond issue. If the credit rating is not affected by the issuance of this obligation, it should be considered a strong option in that it is a cost-effective financing vehicle. However, if the double-barrel bond adversely impacts the general obligation rating, then it might be more appropriate to use another type of revenue bond option. Also, if voters disapprove of the issue, other financing methods not contingent upon voters' approval, such as a revenue bond, will have to be considered.

4. MORAL OBLIGATION DEBT

Moral obligation debt is usually defined as a promise to pay under default debt service of an agency or authority that issues debt. For example, participating cities and counties may have a moral obligation to replenish the debt service fund of a utility issuing revenue bonds to build facilities serving the city and county. Because of strong commitments by cities and counties to water and wastewater utilities serving their jurisdiction, rating agencies view a moral obligation as a credit enhancement. As a result, a government agency may be able to enhance or strengthen its rating on a bond by the use of a moral obligation debt. In addition, moral obligation debt does not require voter approval. This type of debt can also improve marketability of bonds due to the moral obligation pledge. Moral obligation debt does not dilute the strength of the general obligation pledge, but enjoys some of its security.

Moral obligation debt, however, is not an unconditional obligation on the part of a participating government entity. The obligation is subject to approval by a political body, thus introducing a political component into the financing. In addition, it is usually rated one-half to one grade below a general obligation credit rating, thereby increasing interest cost as compared to general obligation debt.

Table 3.2 summarizes the advantages and disadvantages of alternative long-term financing techniques.

E. Innovative Long-Term Financing Methods

In response to high long-term interest rates and an uncertain market, innovative government issuers in the early '80s developed ways to transfer some of the benefits of short-term borrowing to the financing of long-term infrastructure projects. As a result, a number of financing instruments were structured on a long-term basis but priced on short-term periods or puts. Many of these long-term financing instruments remain popular today. With these instruments, the issuer is, in effect, giving up certain financial benefits for lower initial interest rates. Figure 3.1 depicts the trade-off by the issuer of receiving lower initial interest rates for providing credit enhancements such as specific redemption and conversion provisions, allowing interest rates to fluctuate, and providing credit facilities such as a line of credit, letters of credit and bond insurance.

A number of innovative long-term financing opportunities are available in the marketplace:

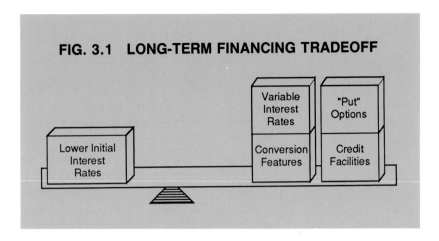

FIG. 3.1 LONG-TERM FINANCING TRADEOFF

- variable rate demand bonds/obligations
- adjustable rate bonds
- zero coupon bonds
- interest rate payment exchange agreements (swaps)
- certificates of participation
- leasing

These innovative financing vehicles are discussed below.

1. VARIABLE RATE DEMAND BONDS/OBLIGATIONS

Variable rate demand bonds offer an attractive low-cost floating interest rate with a long-term maturity. An interest "reset" period is established by the issuer with the interest rate based either on an appropriate index (such as the 30-day exempt commercial paper rate, the T-bill rate, or the London InterBank Offering Rate), or on the rate required to remarket all "put" securities (i.e., totally market-driven). At each "reset" period, the interest rate is adjusted accordingly.

Bond holders typically are allowed to "redeem" their bonds on some predetermined basis. Notice requirements could be as short as daily. If the bonds are "put" (redeemed by the investor), they are usually remarketed immediately. If market conditions are favorable, VRDBs could be converted to a fixed rate of interest for the remaining term of all the bonds.

VRDBs have become popular with investors as well as issuers. Issuers typically receive favorable interest rates in exchange for the put feature and frequent interest rate adjustments. The put feature is

Table 3.2   Conventional Long-Term Financing Instruments

Type Instrument	Advantages	Disadvantages
General Obligation Bonds	• Bonds have lowest interest rates available on a long-term fixed rate bond. • Reserve funds are not typically required. • Administration issuance of bonds is relatively simple, and issuance costs are typically less than other types of debt. • Bonds provide additional security during the construction stage when operating revenues have not materialized.	• Voter approval is usually required for bonds; as a result, long delays before issuance could result and there is no guarantee that voters will approve the issue. • Bonds dilute debt capacity of issuing entity. • Competitive bids are typically required, which can delay issuance.
Revenue Bonds	• Bonds do not affect the debt capacity of the issuing agency. • Bonds can be used in certain situations where general obligation debt is unavailable. • Voter approval is not normally required. • Market timing is more flexible. • Bonds can be presold by underwriter to reduce risk and hedge against market volatility.	• Issuance costs tend to be higher than with general obligation debt. • Investors require higher interest rates with revenue bond issues. • Bonds carry greater risk of default due to the uncertainty of the revenue stream. • Restrictive indentures require special reserve funds to be established. • Coverage requirements on debt service and additional bond tests are typically a part of the bond indenture. • Since these bonds usually require additional financial restrictions that translate into higher debt service, they usually have a greater impact on the user charges.

Table 3.2   Continued

Type Instrument	Advantages	Disadvantages
Double-Barrel Bonds	• Bonds have same advantages as general obligation bonds except debt service payments are secured first by project revenues and second by the taxing power of the issuing agency.	• Bonds have same disadvantages as general obligation bonds except that issuance of the bonds may not have as much of a negative impact on bond capacity and credit rating as general obligation bonds.
Moral Obligation Bonds	• Interest rates are typically better than with revenue bonds. • Bonds do not require voter approval. • Moral obligation feature can improve marketability of bonds. • Bonds do not dilute general obligation pledge but can enjoy some of its security.	• Pledge is not legally binding by agency pledging its moral obligation. • Interest rates are typically above general obligation rates. • Moral obligation pledge must be approved through the political process.

typically backed by a credit facility. If the issuing institution is financially sound, VRDBs can be sold typically with a liquidity facility or line of credit. If credit enhancements are required by the issuing institution, then a letter of credit may be required. Normally, VRDBs sell in minimum denominations of $50,000. Interest is usually paid on a monthly or quarterly basis. In addition, some VRDBs are placed privately, thus reducing issuance cost. In some states, variable rate financing is not allowed without special legislation.

2. ADJUSTABLE RATE BONDS

Adjustable rate bonds are similar to VRDBs in that they provide issuers with floating rates with long-term maturities. Their adjustable rate requirements and put flexibility, however, tend to be more restrictive than with VRDBs. The interest rate for ARBs is adjusted at predetermined intervals, ranging from three months to five years. The bond holders are also entitled to put their bonds at the same time that interest rates are adjusted. ARBs can also be converted at predetermined periods to long-term fixed rate bonds at the issuer's discretion. Because the put options, interest rate adjustment periods, and convertible features are more restricted than VRDBs, ARB interest rates tend to be somewhat higher than rates for VRDBs. ARBs tend to be purchased primarily by institutional investors and issued in denominations of $5,000. Interest is typically paid semiannually. In addition, ARBs are normally secured with a credit facility. Like VRDBs, however, some states restrict the use of ARBs.

3. ZERO COUPON BONDS

Another development in the tax-exempt bond market is zero coupon bonds or capital appreciation bonds. These securities are typically issued at a fraction of their maturity value, with interest on the bonds being added to the value of the bond, as opposed to being paid periodically to the investor. The attractive feature of tax-free zero coupon bonds is that capital gains are accrued with the bond issue until the maturity date, at which time taxes are paid on the interest earned during the holding period. In other words, a zero coupon bond is purchased at a discount and matures at full value, the difference between the purchase price and maturity value being the interest earned over the life of the bond. As a result, instead of receiving tax-free interest periodically during the amortization period, the investor receives a lump-sum, tax-free gain at the maturity date of the bonds.

A major advantage to the issuer of zero coupon bonds is that no payments on the bonds are required until maturity. An issuer may want to provide, however, a sinking fund for retirement of the bonds at some future date. Zero coupon bonds are attractive to the investor in periods of falling interest rates, since the interest rate is based upon market conditions at the time of issuance. A zero coupon bond can be considered like a retirement annuity with the added benefit of tax-free proceeds, unlike either an individual retirement account or 401K plan. Table 3.3 summarizes the advantages and disadvantages of variable rate bonds, adjustable rate bonds, and zero coupon bonds.

4. INTEREST RATE PAYMENT EXCHANGE AGREEMENTS (SWAPS)

An interest rate swap is a contractual agreement between two parties — generally the issuer and an investment bank — to exchange interest payments for a set period of time. The payments are based on a "notional" amount and are only for interest on that amount; no exchange or loan of principal is involved. One party to the contract agrees to pay the other party at a fixed interest rate and to receive variable rate payments based on an index (such as the J.J. Kenny Tax-Exempt Index); the other party agrees to pay variable rate payments and receive fixed interest.

An interest rate swap can be used to achieve financing objectives at a lower cost. For example, instead of issuing short-term notes or variable rate obligations in anticipation of a future long-term fixed rate bond issue, the issuer can immediately sell the long-term bonds and enter into a "variable payor" swap contract for the construction period. The issuer would have the equivalent of variable rate construction period financing without the risks of two-stage or convertible bond financing. Conversely, an issuer with outstanding variable rate debt would stabilize interest payments by becoming a "fixed payor" party to a swap contract. The issuer would pay interest at a fixed rate and receive interest at a variable rate, effectively cancelling out the interest obligation on its outstanding variable rate debt.

5. LEASING

Under lease financing, one party owns a capital item and leases it to a "using" party. The party owning the capital item is called a lessor, and this party leases the item to a lessee. The lessor retains tax advantages of asset ownership with the lessee paying a rent or lease payment

Table 3.3   Innovative Long-Term Financing Instruments

Type Instrument	Advantages	Disadvantages
Variable Rate Demand Bonds/Obligations (VRDBs/VRDOs)	• Bonds provide one of the lowest interest rates available in the long-term financing market. • Financing is available for extended periods (up to 35 years). • Bonds can be converted to fixed rates or other short-term notes.	• Bonds can usually be put back to the issuer on interest rate adjustment dates. • Interest rate adjustment period is frequent (anywhere from 1 to 30 days). • Line or letter of credit is typically required. • Some states disallow the use of VRDBs.
Adjustable Rate Bonds (ARBs)	• Interest rates are typically lower than long-term fixed rate bonds. • Interest rate adjustment period and "put" period are typically longer than with VRDOs. • Bonds can be issued in small denominations ($5,000)	• Interest rates are higher than VRDBs. • Line of credit is normally required. • Some states disallow the use of ARBs.
Zero Coupon Bonds	• Debt service (interest and principal) can be delayed until maturity. • Tax-free capital gains for the investor can be accumulated and used at maturity.	• Interest may be higher than other long-term debt instruments. • In a declining interest rate environment, rates may be unfavorable for an extended period. • Failure to set up appropriate sinking fund could require large outlay by issuer at maturity.

to the lessor. In some cases, there is a purchase provision in the lease which allows the lessee to purchase the capital item during or at the end of the lease period at some predetermined amount, based on an agreed-on formula. For all practical purposes, the lessee uses the capital item as if it were the lessee's asset.

Infrastructure leases for such assets as water or wastewater treatment plants are almost like bond issues. They have predetermined lease amounts for lease periods up to 20 years. Interest rates on tax-free leases usually are slightly higher than revenue bond rates. Advantages of leases include the ease of issuance and administration. To determine the economic feasibility of a lease, a discounted cash flow or net present value analysis should be performed to contrast this technique with other alternatives.

F. Other Issues

A utility issuer should carefully examine what debt instruments are available to it in selecting the most appropriate financing plan. Care should be taken to consider all state laws that might restrict the use of certain types of debt, the timing of cash requirements by the utility and the impact on utility customers, the risk associated with each instrument, and the overall financing goals of the utility and objectives of the community.

At the time of this writing, the U.S. Supreme Court had ruled that Congress has the authority to remove the tax-exempt status from general obligation bonds, revenue bonds, and other tax-exempt financing instruments. There is no indication at this point that Congress will exercise this authority. In the present and future climate of tax reform, however, Congress may look to remove this tax-exempt status as a way of raising federal revenue and reducing the federal deficit. As a result, Congress' potential actions related to tax-exempt status of financing instruments should be considered in determining the timing of the use of financing instruments.

4   CAPITAL RECOVERY CHARGES

4   CAPITAL RECOVERY CHARGES

Capital recovery charges, also known as "system development fees," "facility expansion charges," "utility expansion charges," "capacity charges," and "impact fees," have been increasingly relied upon by governmental utilities as a source of financing for capital improvements. Capital recovery charges are generally established as one-time charges assessed against developers or new water or wastewater customers as a way to recover a part or all of the cost of additional system capacity constructed for their use. Their application has generally occurred in areas that are experiencing extensive new residential and/or commercial development. Public sentiment in certain rapidly growing jurisdictions has shifted from funding all infrastructure costs through increases in user charges to other innovative and creative ways of financing, such as capital recovery charges.

Capital recovery charges have become popular because they attempt to add equity to the financing and pricing system. New capital required for expanding a water and wastewater system to accommodate growth is usually more costly per unit of capacity than historical capital costs for facilities benefiting existing customers. If bonds were issued to finance the full cost of expansion facilities, debt service payments recovered through rates could cause user charges to increase significantly.

Capital recovery charges theoretically add equity to the pricing system by requiring the new customer to make up-front contributions so that rates are not increased to finance expansion facilities. In essence, existing customers would not have to subsidize growth. Capital recovery charges usually recover costs associated only with the major capital components of a water or wastewater system. For a water system, these facilities would normally relate to source of supply, transmission, treatment, and major pumping components. For a wastewater system, treatment, transmission, and disposal facilities would typically be considered. Facilities related to local service lines, water and sewer tap installations, and other facilities benefiting a specific customer or development would normally be recovered through other charges such as developer contributions, water and sewer tap fees, and assessments.

A. Objectives of an Effective Capital Recovery Charge Structure

A community has significant flexibility as to how it establishes a capital recovery charge structure. Determining an appropriate capital recovery charge structure is a complex process requiring several policy and technical issues to be addressed:

- What methodology should be used in determining the charge?

- What percentage of expansion costs should be recovered from new and existing customers?

- What facilities (supply, pumping, treatment, transmission, distribution, collection, disposal, administrative, etc.) should be used in calculating the fees?

- How should grants, developer contributions, and other "contributions in aid of construction" be considered?

- What measure (meter size, equivalent residential unit, etc.) should be used in assessing the charge?

- What period of time should be used for determining the charge?

- Should offsets to the charge be considered for capital costs that will be paid by new customers through their user charges?

- Should the charge be established on a system-wide basis, or should there be separate charges for specific service districts?

In evaluating alternative capital recovery charge structures that address the questions above, several pricing criteria should be considered:

- *Equity:* Is the fee equitable, in that it recovers the cost fairly from the beneficiaries of the service?

- *Revenue potential:* Does the fee generate sufficient revenues to address the community's capital requirements?

- *Potential for litigation:* How likely is it that the capital recovery charge structure will result in litigation from developers or new water and wastewater users?

- *Implementation:* How difficult is it to implement the charge structure?

- *Simplicity:* How easy is the charge structure to explain and to update in future years?

- *Legality:* Does the capital recovery charge structure comply with the appropriate local, state, and federal requirements?

- *Impact on economic development:* Does the implementation of a charge structure unfairly burden growth?

As can easily be seen, many of these criteria are conflicting. As a result, alternative capital recovery charge structures should be carefully evaluated to achieve the optimal trade-offs among these pricing criteria.

B. Advantages and Disadvantages of Capital Recovery Charges

Assessing capital recovery charges upon new developments or new customers has been a source of controversy and legal challenges. From a community's standpoint, these charges offer a number of advantages:

- The charges are paid up-front to the utility, enabling the community to provide additional services immediately. The revenue is worth more paid in a lump sum than if it were paid over time to the community.

- Usually, they are administratively easy to collect.

- Unlike certain types of bonds, the charges do not normally require a vote of the public.

- The charges are an equalization device. They require new development or customers to "buy into" the city's public infrastructure at a fair rate and to repay users who have subsidized the system-wide facilities through prior service charges or taxes.

- The charges impose the cost of extra capacity for infrastructure facilities upon properties that create the need for those facilities.

- The charges provide an additional source of revenue to bolster otherwise inadequate funds for constructing and/ or maintaining essential facilities and services. As a result, less pressure is placed on taxes and user charges for financing capital items.

However, homebuilders, land developers, and new water and wastewater customers object to capital recovery charges because:

- Capital recovery charges add to the "front-end" cost of housing, making new housing less affordable to low- and middle-income families.

- While new home buyers should be required to bear the costs of facilities and other improvements that benefit them directly, capital recovery charges represent a subsidy of preexisting services and are particularly unfair to new home buyers who are long-standing community residents.

- Capital recovery charges are not deductible for federal income tax purposes, making them more expensive in relative terms than ad valorem taxes, if taxes are used to retire debt for infrastructure facilities.

C. Methods of Calculating Capital Recovery Charges

Numerous approaches to developing capital recovery charges have been adopted by government utilities across the country. The major goal in selecting a capital recovery charge methodology is to select an approach which provides equity to existing and future customers and is legally defensible. Several capital cost recovery approaches that are prevalent are discussed below.

1. GROWTH-RELATED COST ALLOCATION METHOD

This methodology embraces the philosophy that capital recovery charges should relate to specific facilities that are designed to accommodate growth. Under this method, a projection of growth-related capital improvements currently under construction or projected to be involved during a projection period (usually five to ten years) is first determined. The number of units to be served by these improvements is then estimated. The unit charge is derived by dividing the costs of growth-related improvements by the number of projected units to be served over the projection period. In some cases, the cost of excess capacity in existing facilities is considered in calculating the fee.

2. MARGINAL-INCREMENTAL COST APPROACH

This method is based on the economic principle that new system users should be responsible for the next increment of capital cost which they cause to be incurred. Capital recovery charges would be designed to recover the cost of this expansion using recent construction cost experience or estimated cost of future facilities. Under this approach, a capital recovery charge would be designed so that existing customer rates would not have to be increased over the planning period.

3. SYSTEM BUY-IN METHODOLOGY

Under this approach, capital recovery charges are based upon the "buy-in" concept that existing users, through service charges, tax contributions, and other up-front charges, have developed a valuable public capital facility. The charge to users is designed to recognize the current value of providing the capacity necessary to serve additional users. The charge is computed by establishing fixed asset value under a historical or reproduction cost basis and deducting relevant liabilities (long-term debt, loans, etc.) from this amount. The number of units of service is then divided into this difference (considered to be the utility's equity) to establish the capital recovery charge.

4. VALUE-OF-SERVICE METHODOLOGY

Under this approach, capital recovery charges are based on the practices of similar communities, tempered by the perceived ability of new users to pay. In general terms, the price structure is tied to the

Table 4.1   Meter Demand Ratios

Meter Size	Ratio of Demand
5/8″	1.0
3/4″	1.1
1″	1.4
1 1/2″	1.8
2″	2.9
3″	11.0
4″	14.0
6″	21.0
8″	29.0

Source: American Water Works Association (AWWA) Manual M-1.

concept of "what the market will be." A community should be cautious in using this method, however, in that the fee is not directly related to cost of service, and litigation could result from excessive fees. A simple example of how to calculate capital recovery charges under each methodology is provided in Appendix A.

D. Assessing Capital Recovery Charges

To facilitate public understanding and administration, capital recovery charges are often structured around a readily determined standard. This standard attempts to differentiate among customers based upon the loading that different customers or customer classes place on the system. Common ways of assessing a capital recovery charge include (1) meter size, (2) equivalent residential units (ERUs), and (3) drainage fixture units (DFUs) or supply fixture units (SFUs). Each of these methods implies different levels of equity, understandability, complexity in calculating, and difficulty in implementing.

1. METER SIZE APPROACH

Under the meter size approach, a customer's meter size is used as the basis for assessing the capital recovery charge. As demonstrated in Table 4.1, the charge usually varies based on the ratio of potential demand of the five-eighths-inch meter to other water meter sizes.

Once the charge for a five-eighths-inch meter is determined, other meter sizes are calculated by multiplying their relevant ratio by the five-eighths-inch meter charge.

The major disadvantage of the meter size approach is that it normally does not differentiate enough among users within the same meter class. In other words, there could be major differences in loading among customers in the same meter class. For example, a one-person household with a five-eighths-inch meter using 3,000 gallons of water per month puts a significantly lower loading on the system than an office with a five-eighths-inch meter using 20,000 gallons per month. Yet under the meter size approach, both customers would pay the same capital recovery charge.

2. EQUIVALENT RESIDENTIAL UNIT APPROACH

The equivalent residential unit approach attempts to correct for some of the inequities of the meter size approach. Under the ERU approach, utility customers are classified by common business or residential characteristics. The characteristics used attempt to place a customer into a common loading category or measurement system. For example, residential customers could be categorized into single-family dwellings, townhouses, duplexes, mobile homes, and apartments. More precise differentiation could be made within each residential category based upon number of bathrooms, number of bedrooms, or square footage. Commercial establishments could be differentiated by the nature of the business. Examples of commercial ERU categories would be restaurants, car washes, service stations, or other commercial groupings. Within each business category, some common measure to reflect loading would be established. For example, common measures such as square footage, number of seats (restaurant), and number of rooms could be used.

Even though additional equity can be achieved under the ERU approach, it can be cumbersome. In addition, some customers argue that this approach still does not take their unique loading characteristics into consideration.

3. FIXTURE UNIT APPROACH

The drainage fixture unit and supply fixture unit approaches attempt to provide additional equity in assessing capital recovery charges. Under this approach, specific fixture units are identified for each new customer connecting to the water or wastewater system. Using a generally accepted plumbing code, the number of gallons of loading is determined for each type of fixture (toilet, washing machine outlet, faucet, etc.). A customer's capital recovery charge is then determined by multiplying the number of fixture units by the capital recovery charge (expressed in fixture units). In essence, each customer has an individual capital recovery charge calculated based upon his precise number of fixture units. The obvious disadvantage of this method is that it requires an analysis of the fixture units for each new customer, which makes it more complicated and administratively difficult to implement. The advantages and disadvantages of the three assessment methods are contrasted in Table 4.2.

E. Other Considerations Related to Developing and Implementing Capital Recovery Charges

1. LEGAL ISSUES RELATED TO CAPITAL RECOVERY CHARGES

Capital recovery charges are generally established by local ordinance. Several states, however, have statutes which limit and/or regulate the implementation of capital recovery charges. For example, the *Dunedin* case in Florida defines how capital recovery charges should be calculated and how funds should be used. Care should be taken to insure that the development, method of assessment, and implementation of capital recovery charges comply with all local, state, and federal requirements.

2. WHEN TO ASSESS THE CAPITAL RECOVERY CHARGE

Many government utilities require payment of the charge (1) immediately upon receipt of an application or a building permit, or (2) at the time a user receives a particular service from a community. Most capital recovery charge structures make the land developer or homebuilder responsible for paying the charge, rather than the ultimate home buyer or commercial property owner. If the developer or homebuilder pays the charge, it is likely added to the price of the house or commercial property. The timing of the collection of the charge has a direct effect on development costs, since the developer will often rely on interim financing to pay the charge.

3. USE OF CAPITAL RECOVERY CHARGE FUNDS

Capital recovery charges are primarily used for the purpose of financing capital facilities associated with growth and expansion. In such cases, it is important that capital recovery charges be restricted

Table 4.2 Methods of Assessing Capital Recovery Charges: Comparison of Advantages and Disadvantages

Assessment Method	Advantages	Disadvantages
Meter Size	• Easily understood by customers. • Generally easy to implement. • Encourages proper meter sizing by customers (larger meters would require higher capital recovery charges). • The customer's meter size identifies the potential demand placed on system by the customer.	• Significant loading differences can exist among customers within a specific meter size class. • For wastewater, meter size may have little relevance to the customer's wastewater loading.
Equivalent Residential Units	• Attempts to recognize loading differences among different classes of customers—provides additional equity when compared with the meter size approach. • Can be used effectively for both water and wastewater capital recovery charges.	• Bases the charge on usage class characteristics rather than the precise plumbing configuration of the customer. • Is more difficult to explain and administer than meter size approach.
Drainage Fixture or Supply Fixture Units	• Generally provides the most equitable capital recovery charge in that the unique plumbing configuration of the customer is used. • Can be effectively used for both water and wastewater capital recovery charges.	• Can be the most costly and administratively difficult to implement in that each customer's building plan must be individually reviewed.

for expansion purposes and not used to finance capital costs that should be shared by all users of various facilities. Special accounting procedures need to be established to insure that capital recovery charges are segregated and proper reporting made.

4. CREDIT FOR DEBT SERVICE

Many times, bonds are issued to finance major facilities for expansion. Debt service on bonds is then recovered through either user charges or taxes. If the new homeowner is required to pay a capital recovery charge on facilities that are bond financed, he argues that he is charged twice — once when the charge is paid, and a second time when he pays his taxes or user charges to retire debt. To address this inequity, a "debt service credit" is made to the capital recovery charge to offset amounts the new homeowner will pay through taxes and user charges to retire debt.

In summary, capital recovery charges can be a powerful and equitable method for financing new capital facilities for utility and non-utility purposes. Care should be taken, however, to insure that the capital recovery charge structure addresses the planning goals of the community. Many complex factors have to be considered in developing the most appropriate capital recovery charge structure for a particular community. Adequate time should be allowed (1) for collection of appropriate data to develop the capital recovery charge structure, (2) to ensure that the methodology chosen is responsive to the planning goals of the community, and (3) to ensure that proper input is received from the community and government officials in developing the charge structure.

F. Comparison of Capital Recovery Charges

Arthur Young's 1988 National Water and Wastewater Rate Survey compares one-time connection charges of major governmental water and wastewater utilities across the country. The survey focuses on what charge a typical single-family customer would pay to connect into a water and wastewater system. In many of the communities surveyed, the major part of the connection charge tends to be the capital recovery charge. In other communities, however, a large part of the connection fee is related to meter installations, water and sewer tap costs, cost of local service, and other related costs. With this in mind, the results of the survey are presented in Appendix B.

5   PRIVATIZATION

5   PRIVATIZATION

Over the past several years, government utilities have become more conscious of providing public services in an environment of increasing cost, growing demand for services, and an insistence by the public to limit tax and user-charge increases. In the early and mid-1980s, several factors contributed to the emergence of privatization as an attractive alternative to traditional methods of providing services to the public. First of all, federal and state grant funding providing monies for financing public infrastructure facilities declined significantly over the last six years. Washington determined that in order to control the federal deficit, major restrictions on spending were required. As a result, fewer dollars were available to local communities for financing much-needed improvements. At the same time, tax laws were passed to make private ownership of certain capital facilities very attractive. The Economic Recovery Act of 1981 was the first major tax act to encourage capital investment by private investors. Tax law amendments in 1982 and 1984 specified conditions and constraints on leasing and privatizing activities, but still provided a means by which the private sector could profitably enter into service relationship with public entities. Even with the passage of the Economic Recovery Act of 1986 and the Deficit Reduction Act of 1987, several tax advantages of privatization still exist for both the public and private sectors.

A. Overview of Privatization

Privatization can be defined as: *private sector involvement in the design (if appropriate), financing, construction, ownership, and/or operation of a facility which will provide services to the public sector.*

FIG. 5.1.   THE PRIVATIZATION "WIN/WIN" SCENARIO

To Work for the Private Partner:

Service Charge Revenue
    Plus
Revenues from Sale of Facility
    Less
Operating and Maintenance Costs
    Less
Debt Payments under Privatization
    Plus
Tax Benefits

$>$ Return on
Alternative Investment
(Cost of Capital)

To Work for the Public Partner:

Service Charges
    Plus
Purchase Payment

$<$ Operating and
Maintenance Cost
& Debt Payments
Under Public Ownership

Privatization of water and wastewater facilities is a way for the private sector to work with local governments in obtaining and/or operating needed facilities. Privatization can take several forms, ranging from "contracting out" to private financing and ownership of facilities and providing of service through a service contract. In keeping with the theme of the book, this chapter focuses primarily on private financing and ownership of facilities.

These public/private partnerships are based on the sharing of benefits. In a typical privatization transaction as discussed in this chapter, the private sector receives the business opportunity of owning and operating a water or wastewater treatment plant, and the local government receives cost-effective delivery of a necessary service. Figure 5.1 depicts the financial balance of how a privatization transaction works. Privatization may be used for all types of water and wastewater facilities, including collection systems and facilities, conventional and recycle/reuse treatment plants, and sludge facilities. Of these, however, the most attractive candidates for privatization are typically stand-alone facilities that are either a system in and of themselves or that are completely separated from the other portions of the system.

FIG. 5.2. PRIVATIZATION ARRANGEMENT

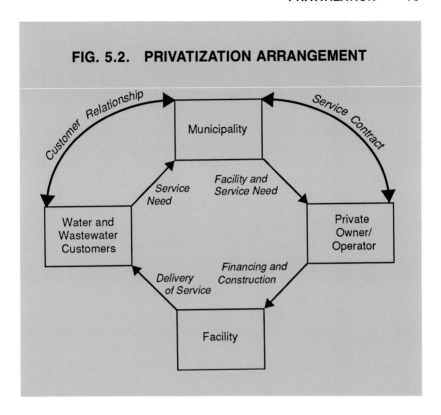

B. The Structure of a Privatization Transaction

The typical privatization transaction is shown in Figure 5.2. The relationships between the parties involved in the project are summarized as follows:

- The private company arranges the financing, constructs and operates the facility, and delivers the service to the water and wastewater customers.

- The governmental water and wastewater utility pays the company a set fee for supplying the service.

- The customers pay the utility for the service, just as though the utility was providing the service. The customers still maintain their relationship with the utility as if the utility owned and operated the facility.

The key aspect of privatization is that it is a partnership between the public and private sectors. This partnership is embodied in a legal document typically referred to as a service contract. The service contract is important, since such a contract must exist for a privatization transaction to be successful. Key provisions to be incorporated into the service contract are presented in Figure 5.3.

Several communities have chosen privatization as the way to finance, construct, and operate key water and wastewater facilities in their service area. Examples of these communities include Pelham and Auburn in Alabama, and Chandler and Gilbert in Arizona.

C. Why Privatization Works

The attractiveness of privatization lies in economic as well as non-economic benefits. Under privatization, the potential exists for the private sector to apply construction time and cost savings, efficient operations, and tax benefits to lower the costs of environmental services and share these savings with the public sector in the form of lower user fees.

The potential savings on a project will vary according to site-specific circumstances. Even in cases where the savings are minimal, privatization may be appropriate due to limitations on local debt capacity, the need for more timely delivery of services, or other advantages perceived by local officials. The advantages are discussed further below.

1. CONSTRUCTION SAVINGS

Construction savings may result from the private sector's ability to procure materials and proceed through the design and construction at a much faster rate than can the government utility. As discussed below, the community benefits from more predictable costs because the community has a contractual commitment from the private sector partner in the transaction as to service provision costs.

2. PROCUREMENT AND SCHEDULING

In any major construction project, time savings has the potential of being translated into cost savings. Many industry observers regard the delays associated with the procedural requirements and approval processes of public works construction as major factors in increasing the

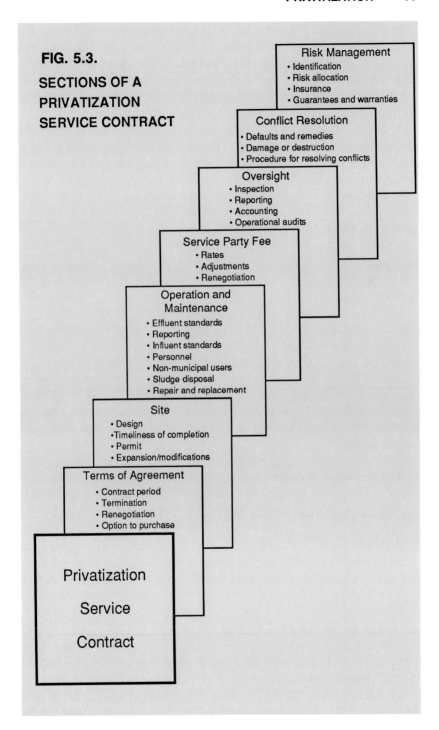

FIG. 5.3.

SECTIONS OF A PRIVATIZATION SERVICE CONTRACT

Risk Management
• Identification
• Risk allocation
• Insurance
• Guarantees and warranties

Conflict Resolution
• Defaults and remedies
• Damage or destruction
• Procedure for resolving conflicts

Oversight
• Inspection
• Reporting
• Accounting
• Operational audits

Service Party Fee
• Rates
• Adjustments
• Renegotiation

Operation and Maintenance
• Effluent standards
• Reporting
• Influent standards
• Personnel
• Non-municipal users
• Sludge disposal
• Repair and replacement

Site
• Design
• Timeliness of completion
• Permit
• Expansion/modifications

Terms of Agreement
• Contract period
• Termination
• Renegotiation
• Option to purchase

Privatization Service Contract

cost of public works construction, even though the government utility uses the competitive low-bid technique of procurement. Under a public sector approach, every major decision has to be made by the community. The framework and procedures for decisionmaking may vary greatly from community to community, but by and large, they are time-consuming. In contrast, the majority of decisions can be made by the private sector rapidly and without delays. The potential for savings of time should be reflected in a lower overall cost of construction, which should contribute to the ability of the private sector to reduce fees for service.

The community is not totally removed from the process, but continues to be involved only to the extent necessary. The community and/or its technical advisors would be involved until the design is completed and can be recommended to the community for acceptance.

3. RISK REDUCTION

In a properly structured privatization approach, communities should face fewer design, construction, and operation risks. Under a municipal approach, the potential does exist for design errors and omissions which, if experienced during construction and/or operation, can cost the community additional money, time, and headaches. Under privatization, the private firm will warrant the design and be responsible for any increased costs due to errors and omissions under their control.

During construction, there is also the potential for construction defects and failures, cost overruns, and claims. Under privatization, the private firm bears the risks related to construction because it bears the burden of change orders, claims, cost overruns, etc.

4. OPERATIONAL SAVINGS

The private sector can usually more readily accommodate the pay scales necessary to attract and retain highly technical and competent individuals. The talents of senior or specialized individuals can also serve multiple facilities. Through the economies of scale realized when a private firm operates multiple facilities, the system's users can also realize savings when the private sector pursues initiatives such as centralizing administrative staffs and systems, ordering supplies in bulk, and sharing common inventory items, operators, and maintenance personnel among several facilities. In addition, operational cost savings can be realized through facility design. Since the privatizer/designer will become the owner/operator, operating costs will likely be more closely scrutinized.

5. TAX BENEFITS

In a privatization transaction, the private sector firm is entitled to certain tax benefits from which local municipal governments do not benefit. The tax benefits associated with privatization, and provided for under the Tax Acts of 1982, 1983, and 1984, included Accelerated Cost Recovery System (ACRS) depreciation, an Investment Tax Credit (ITC), and deductibility of interest on the debt associated with the project. Several of the tax benefits were repealed or modified under 1986 tax legislation. The Deficit Reduction Act of 1987 eliminated ITC and significantly reduced the use of accelerated depreciation. Tax benefits still exist, however, in the form of depreciation and interest-expense deductions. These deductions are not as attractive, however, as they were before tax reform.

Another previous tax benefit was the ability to finance privatization transactions with tax-exempt debt utilizing industrial development bonds (IDBs). The use of tax-exempt debt equalized the "cost of money" for the privatization transaction and the "cost of money" for a municipal government through general obligation or revenue bonds. In other words, interest rates on the debt were by and large equal. Again, with the passage of the Deficit Reduction Act of 1987, the use of tax-free debt for such programs has been significantly restricted. In many states, however, the private sector may be able to use innovative financing that is not available by statute to local government utilities.

6. DEBT CAPACITY BENEFITS

In many states, privatization does not require a community to encumber its debt capacity even if, as in rare instances, IDBs are used to finance the project costs. Because IDBs are backed by the revenues of the project, and therefore are not considered a general debt of the community, the community can retain its debt capacity for other essential services. In time, however, rating agencies may consider the community's long-term commitment in the service contract between the community and the private firm as an obligation which could affect the community's ability to borrow.

7. AVAILABILITY OF FINANCING

In some circumstances, privatization may be the only alternative available to enable a community to meet its treatment needs on a timely basis. Many communities have little or no chance to obtain a federal grant or state aid for needed facilities. Some eligible communities may face delays of several years before receiving grant funds or state aid. In these instances, there may be advantages to proceeding with the private approach immediately. If the delay is substantial (i.e., a few years), the increase in construction costs due to inflation may offset the advantage of grant funding or state aid. Add to this the fact that the level of funding or aid is uncertain and dependent upon the eligibility requirements, and the grant or aid approach may lose attractiveness. It may make sense for a community to assess the feasibility of proceeding immediately with privatization in view of the risks associated with waiting for federal grant funding or state aid.

Another advantage of privatization relates to "incomplete credits." For example, the credit worthiness of a project in a poorly rated community may be improved (and financing cost reduced) when a strong privatizer is involved, providing cost and performance guarantees.

Finally, the economic attractiveness of privatization is enhanced by the private sector's ability to use financial creativity in the project's financing. Many states limit the public sector's flexibility in using more contemporary forms of financing. Examples of innovative approaches that are not allowed by local governments in some states include variable and adjustable rate financing, use of tax-exempt commercial paper, and various credit enhancements. Even though there is additional risk associated with these more creative financing approaches, more favorable financing could normally result. The private sector's ability to use these financing approaches is not normally constrained.

D. Disadvantages of Privatization

The potential disadvantages resulting from privatization of water and wastewater facilities relate primarily to a loss of control by municipalities, potential negative aspects of a long-term contract, and uncertainties relating to legal and regulatory issues, each of which is discussed below.

1. LOSS OF LOCAL CONTROL

The argument most often heard against privatization is that the municipality loses control over the financing, construction, and operation and maintenance of the facilities. By relinquishing ownership, "hands on" control of the operation and maintenance of the system is lost. That is, even though the municipality can hold the privatizer accountable for performance, on a daily basis the privatizer controls service levels, compliance with treatment standards, discharge levels, etc.

However, the privatization transaction can be structured to provide for local government operation of a privately financed and owned facility. As some of the tax benefits have been taken away under recent tax legislation, public sector operation is many times allowed without endangering remaining tax benefits. In addition, with a privately operated facility the municipality can still retain control in such areas as:

- development and implementation of a user charge system

- primary contact and interaction with users of the facilities

- control of growth within the service area

- responsibility for determining expansion of the facility and under what conditions it can occur

- responsibility for connections and disconnections

- the right to inspect the facilities and the right to perform fiscal, management, and/or operational audits

The local government can specify performance standards and hold the private owner/operator accountable for meeting these standards.

2. NEGATIVE ASPECTS OF A LONG-TERM CONTRACT

Another potential disadvantage of privatization is the reliance on the terms of the service agreement should the transaction "go sour," and the risks and potentially costly litigations that the community will be subject to under such circumstances. Obviously, because of the relative newness of the concept, a historical perspective cannot be taken as to the protection afforded by the service agreement. It is

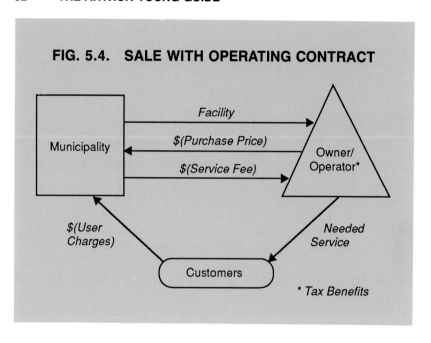

FIG. 5.4.   SALE WITH OPERATING CONTRACT

clear, however, that minimization and proper allocation of risks is of great importance, and it is essential that the community protect itself with qualified legal, financial, and technical advisors to assist in the structuring of the service agreement and overall implementation of the privatization transaction.

E. Several Detailed Approaches to Structuring the Privatization Transaction

There are several specific approaches to privatization. These approaches include a sale with an operating contract, a full-service agreement, a service contract with a third-party investor, and a service contract with an operating company.

1. SALE WITH OPERATING CONTRACT

Under the "sale with operating contract" structure, the government utility sells a facility to a private sector owner who operates the plant under a service contract with the utility (Figure 5.4). The private sector firm is eligible for any tax benefits. This is an appropriate

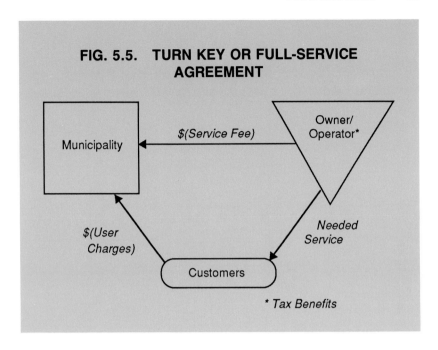

FIG. 5.5. TURN KEY OR FULL-SERVICE AGREEMENT

structure for a facility which has not been federally funded. This structure would be appropriate for a new facility, or for a water or wastewater treatment facility which was built without federal funding. If the facility was built with federal funds, analysis of the feasibility of privatization would have to include consideration of a payback to the federal government of grant funds received.

2. TURN KEY OR FULL-SERVICE AGREEMENT

The "full-service agreement" structure is a traditional full-service operation, where the private sector firm finances, designs, builds, owns, and operates the treatment plant through a coordinated implementation plan with the public sector (Figure 5.5). The community maintains a "retail" relationship with its customers, and purchases services "wholesale" from the privatizer. This is an appropriate structure when a community needs to construct an entirely new facility. The transaction, as illustrated, assumes that one private sector firm is the sole owner of the facility, although variations are possible.

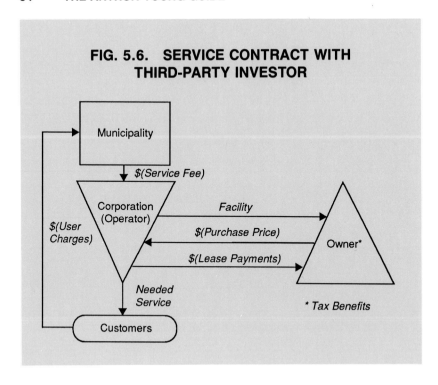

FIG. 5.6.   SERVICE CONTRACT WITH THIRD-PARTY INVESTOR

3. SERVICE CONTRACT WITH THIRD-PARTY INVESTOR

Under the "service contract with third-party investor" structure, the corporation (e.g., an engineering firm) designs and oversees the building of the facility and then enters into a limited partnership with other investors or a form of leverage leasing agreement with a partnership or unrelated financial institution (Figure 5.6). The engineering firm operates the facility under a service contract with the municipality, and the partnership (owner/lessor) receives any tax benefits.

4. SERVICE CONTRACT WITH OPERATING PARTY

Also, the option exists of entering into a service contract with a private firm for the sole purpose of operating the facility. However, under this structure, the only area for cost savings is in operation and maintenance, because the facilities are owned by the municipal government.

F. Determining the Feasibility of Privatization

Assessing the feasibility of the proposed project is imperative and should be accomplished before any other actions are taken. The process of determining the feasibility of privatization is depicted in Figure 5.7 and discussed below.

1. EVALUATION OF TREATMENT NEEDS

The initial step in determining the feasibility consists of a needs assessment in which a clear understanding of the community's current, short-term, and long-term requirements must be addressed. Information available in facilities planning or similar documents, combined with input from knowledgeable local officials, is imperative in this assessment.

2. TECHNOLOGY REVIEW

The community should review technologies that are available to meet its needs. While privatization economics should be secondary to sound engineering, the choice of technology will affect costs in the privatization project. There are, for instance, potentially greater tax benefits available when equipment-intensive technologies are used compared to those potentially available from the use of land-based systems, because land is not depreciable. In addition, the range of available and acceptable technologies must be assessed based upon evaluation of initial and life-cycle costs, benefits, disadvantages (if any), and other related factors.

3. EVALUATION OF VENDOR INTEREST

Vendor interest should be gauged before proceeding too far with the project. Before investing the time, effort, and funds required to issue requests for privatization proposals, a community might seek preliminary expressions of interest from potential privatizers. Strong vendor interest on a preliminary level can enable a community to confidently proceed.

4. ASSESSMENT OF RISKS AND BENEFITS

Once adequate vendor interest is determined, evidence of the risks and benefits of the proposed project should be addressed. The project

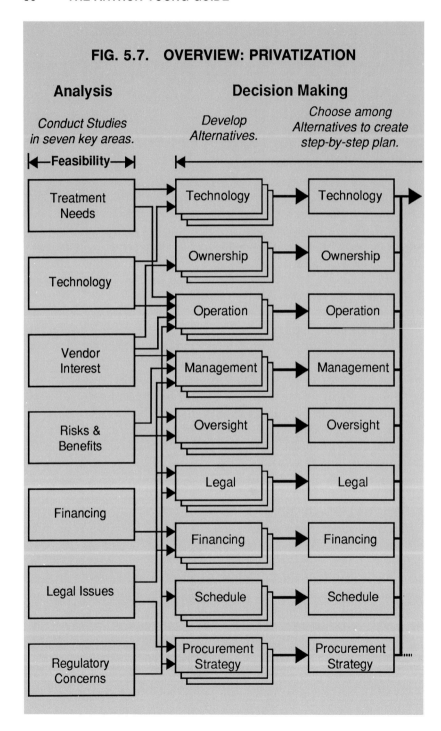

FIG. 5.7. OVERVIEW: PRIVATIZATION

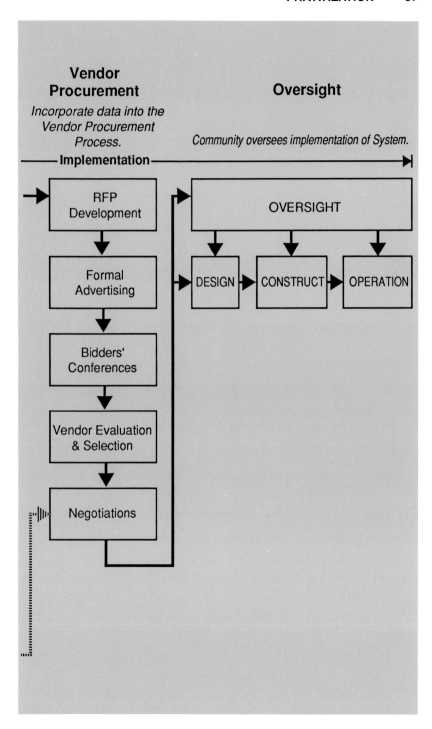

should be assessed on the basis of potential impacts to the political, economic, and social aspects of the community. These factors are likely to include the project's effect upon economic growth and the control of future development. It is equally important to consider educating the public and the local news media about privatization and the partnership concept on which privatization is based.

5. FINANCIAL FEASIBILITY ANALYSIS

The financial alternatives available to fund the project should be analyzed. In this analysis, the economics of privatization are compared to other available financing methods. If privatization is found to be the best alternative, a decision must be made about how to structure the transaction. While it would be appropriate for the community to investigate various financing alternatives and assist as appropriate in the establishment of these options, it would be unwise for the community to unnecessarily limit the financing approaches that may be used by the privatizing firm. Financial creativity of the private sector may lead to lower user fees. A community should limit financing options only to the extent necessary to protect the interests of the users of the facility.

6. EVALUATION OF LEGAL AND INSTITUTIONAL FACTORS

Early in evaluating the feasibility of privatization, the community should address institutional and legal factors that may influence the project. Generally, these legal issues include any existing limits on a community's ability to enter into a long-term contract. The private firms involved in privatization projects typically want assurances that their contracts will last between 20 and 25 years or at least as long as the life of the bond issue used to finance the facility. Other types of institutional issues to consider include potential limitations on a community's method of procurement and possible constraints on the contract between the community and the private company. In some cases, enabling legislation is required to allow privatization projects to be realized.

7. REGULATORY CONSIDERATIONS

Finally, it should be determined what role state government and EPA will assume in regard to compliance with applicable permit requirements, enforcement issues, environmental aesthetics, and other factors. State actions to support privatization are discussed in greater detail later in this section.

In order to adequately address all of the issues involved in assessing the project feasibility, it is recommended that the community conduct a feasibility study. The typical study will determine whether privatization is appropriate to a particular project. It will also identify issues that need to be addressed as the project goes forward, thereby eliminating the potential for problems. Even in situations where an unforeseen issue arises, if a privatization project has been initiated with the assistance of qualified advisors and a thoroughly organized and well-planned approach, the mechanisms will be in place to resolve unexpected issues quickly and conclusively.

G. Privatization Procurement and Implementation

The successful procurement and implementation of privatization will be heavily conditioned on the quality and comprehensiveness of two key documents: (1) the request for proposal (RFP); and (2) the service contract. The overall implementation process for the project will be facilitated if these documents are similar in content, direction, and detail.

1. MODEL REQUEST FOR PROPOSALS

The RFP will be the first exposure that potential privatizers will have to the goals, objectives, and requirements of the community. It will be the potential privatizer's first opportunity to assess the seriousness of the community's intent to implement the project, and will serve as the basis on which they will decide, in view of the community's requirements, whether or not to pursue the privatization opportunity. The RFP will incorporate and clearly delineate all of the provisions, terms, and conditions that are of primary concern to the community. One of the reasons for the comprehensiveness and definition is to elicit proposals that, to the greatest extent possible, can be evaluated on common grounds and will facilitate a smooth and orderly transition to the formulation of the service contract. The outline of a representative RFP is presented in Table 5.1.

Table 5.1   Representative Privatization RFP for Wastewater Treatment Plant

I.   **Introduction**
- Summary of needs and opportunities
- Potential grant status
- Community interest in privatization

II.   **Background Data on the Local Area**
- Local demographics
- Economics
- Population trends
- Climate

III.   **Proposal Format and Requirements**
- Standardized assumptions
- Standardized table of contents
- Standardized bid form

IV.   **Technical Information**
- Treatment needs
- Acceptable technological approaches
- Existing facility and plant site data
- Permit and compliance standards
- Sludge disposal
- Staffing and training
- Engineering services performed to date
- Construction and O&M cost estimates to date
- Engineering documents available for review
- Environmental and aesthetic specifications

V.   **Service Requirements**
- Handling current needs
- Handling future growth
- Procedures to address regulatory requirements; current and potential future modifications
- Service to nonmunicipal users
- Schedule for project completion and startup

VI.   **Legal, Insurance, and Related Matters**
- Customer, labor, and supplier agreements
- Regulatory matters
- Guarantees and warranties
- Governing statutes
- Key contractual considerations
- Methodology to resolve unanticipated events
- Insurance and indemnification requirements
- Damages
- Existing employees
- Term and termination conditions of contract
- Nonperformance penalties
- Schedule delay penalties

Table 5.1 Continued

VII.	**Financial Information**

VII.   **Financial Information**
- Acceptable financing approaches
- Safeguard and transfer of existing investment
- Future investments, guarantees, and warranties
- Community's financial capabilities
- Standardized assumptions
- State and local tax information
- User fee considerations
- Sources of financial assistance
- Facility purchase option

VIII.   **Anticipated Interfaces**
- Community pretreatment management system
- Local community
- Regulatory community
- Community operation of collection system, hook-ups, and disconnects

IX.   **Facility Management System Requirements**
- Construction management
- Oversight program
- Transition procedures
- Access to records and facilities
- O&M management
- Reporting and control systems
- Audit and control requirements

X.   **Risk Allocation**
- Risks expected to be assumed by the owner/operator
- Risks allocated to the community
- Exceptions taken by proposer to community's risk posture

XI.   **Proposal Evaluation Process and Contract Award**
- Evaluation process
- Evaluation guidelines
- Contract award

2. THE SERVICE AGREEMENT

The service contract is an agreement between a local government and a privatizer to provide wastewater treatment services. It is the principal document in any privatization transaction. This contract outlines the terms of the privatization partnership and should include any provisions which protect the interest of both partners in the transaction. The terms of the service agreement will provide principal credit support for the project. Representative sections of a model service agreement are presented in Figure 5.3.

The formal privatization service agreement is an agreement between the community and the private-sector owner of the facilities to provide treatment services. It is the principal document in any privatization transaction. The privatization agreement outlines the terms of the privatization partnership and includes provisions which protect the interests of both parties. The terms of the service agreement will provide the principal credit support for the project. The community establishes rates and collects user fees to secure the revenues necessary for the payments stipulated in the contract.

In the majority of privatization transactions that have been structured thus far, the private sector firm has been the provider of the services to the municipal government. The municipal government makes one payment to the private sector and bills each individual user. Private sector firms expressing interest in privatization favor this method because it is the simplest. The community, however, must recognize that the privatizer may want the right to bill and collect fees from certain nonmunicipal users. This arrangement can also "buffer" the privatizer from state public service commission regulation since the customer ultimately is served by the community, a tax-exempt entity.

For the municipal government, privatization has the added advantage of allowing the local government greater control over the price of treatment services. It also provides the option of adding additional charges, such as existing debt service charges and O&M costs for the local collection and/or conveyance system.

However, it must be determined whether the community has the authority to collect fees under a privatization approach. Some communities may consider collecting fees to pay the charge to the private sector a different issue from collecting user fees for treatment services provided directly by the community. A local ordinance may need to be enacted.

Also indicated in the contract is the rate schedule that the private sector will charge the community. This service charge will consist of two components. One component includes a schedule of fixed charges that represent the monies the private sector firm will use to help in retiring the debt associated with the project. The other schedule will be the O&M costs, which will include rate escalation clauses tied to labor, power and material indices, and "force majeure" (unforeseeable and uncontrollable events or effects). Service charge provisions can also provide for nonroutine adjustments to reflect other events such as new state-imposed O&M procedures or equipment requirements.

A purchase option clause may also be included that will allow the community the option (under very specifically defined terms) to purchase the facility from the private sector. The length of the contract period is another major clause of the contract. In some cases, the private sector will want to maximize tax benefits and will request a contract period of 20 to 25 years. An option to purchase the facilities and a contract renewal option would also be included to cover the period after the initial contract period. The municipality may want to protect its interests by specifying under what conditions it may terminate the contract.

Another major provision is the allocation of liabilities and fines. In most cases, the owner will be responsible for operational differences and the quality of the effluent. However, the private sector's liability will be dependent on the community's meeting its responsibilities as to influent flow and characteristics. Allowable influent quantity and quality should be negotiated and stipulated in the contract.

The issues of the existing work force, labor contract, compensation, benefits, etc. are additional areas for concern and negotiation. Other areas which will necessitate contract clauses include, but are not limited to:

- design and construction oversight, responsibilities, and management practices
- operations and maintenance oversight, responsibilities, and management practices
- handling of residuals
- procedures for unanticipated growth
- environmental standards to be achieved
- enforcement of warranties
- security
- recordkeeping and reporting requirements
- permit, testing, and laboratory requirements
- compliance with federal and state regulations
- O&M expenditures
- staffing
- modifications to existing structures and equipment
- odor control
- training
- municipal responsibilities
- insurance
- municipal access to facility and records
- contract renewal option

- responsibilities under "force majeure" conditions
- transition procedures and responsibilities

H. State Considerations Related to Privatization

In order to maximize the benefits associated with privatization, it is necessary to insure that state laws support the privatization transaction. Many states have undergone studies examining areas to "sweeten" privatization for the privatizer and local government. The typical product of a state study is the drafting of legislation which incorporates changes to state laws to enhance privatization. State enhancements to privatization that could be incorporated in legislation are discussed below.

- *Sales tax exemption:* All private concerns in most states are subject to state sales tax. On the other hand, local governments in most states are exempt from sales tax. Because the privatization transaction can be viewed as a quasi-public entity, it can be argued that the privatizer should enjoy an exemption from sales tax, the savings of which will be shared with the public sector.

- *Property tax exemption:* The discussion relating to sales taxes also applies to property taxes.

- *Conveyance of public property:* Some states disallow the conveyance of public property by a local government to private sector organizations. Conveyance relates particularly to providing sites and any existing facilities that may be included in a project (from buildings to process equipment). This feature would have to be added to state law to insure that the conveyance credential is still allowed.

- *Condemnation of public property for private purposes:* In many states, property can be condemned only for the direct ownership and benefit of a local government. In order to enhance privatization, local governments need to have the power to condemn property on behalf of a privatizer.

- *Regulation by public service commission:* A very restrictive feature in many states is the inability to remove the privatization transaction from control of the public service commission of that state. As a result, rates of return are strictly monitored. To enhance privatization, provisions need to be made in state laws to buffer profitability from regulation by the public service commission. Profitability would best be regulated under the terms of the privatization service contract.

Some of the states that have adopted legislation to enhance privatization include Utah, New Jersey, and Alabama.

Privatization is not an all-encompassing panacea for water and wastewater facility financing and construction. Rather, it is one of several approaches to solve the infrastructural problems facing local government utilities. The availability of federal and state grants and/ or loan programs, the project-specific conditions of a given situation, private sector interests, and public sector acceptance are all factors that, in part, determine the feasibility of privatization. Finally, and not least important, the financial status of a given community and its ability to incur debt at favorable rates can very well point towards 100% local financing for water and wastewater projects as the best course of action.

6

SELECTING THE APPROPRIATE CAPITAL AND FINANCIAL PLAN

6 SELECTING THE APPROPRIATE CAPITAL AND FINANCIAL PLAN

The prior chapters discuss the capital and financial planning process and several capital financing methods. There are numerous combinations of financing methods that could be considered by government utilities selecting an appropriate capital and financial plan. In order to ensure acceptance of the financial plan, the utility should tailor the plan to the financing objectives of the community and its residents. Several of the many economic and noneconomic factors to consider include:

- short-term and long-term interest rates
- time frame and cost of issuance
- risks associated with changing market conditions
- potential tax law changes
- degree of public acceptance of a financing program

These factors and the process of selecting an optional financing program are discussed below.

A. Long-Term and Short-Term Interest Rates

The economic impact on customers is an important issue to consider in selecting an appropriate plan. Since a plan might apply to a planning horizon of 30 to 40 years, the impact on customers during this total period should be considered. One goal of a good financial plan is to match the cash requirements of the plan to the benefit received by users of utility services. Interest rates tend to be the major factor in evaluating the economic feasibility of a particular financing plan. As discussed in Chapter 3, short-term interest rates are generally less than long-term interest rates because the short-term investor has obligated himself to a specific interest rate for a shorter period of time and therefore reduced his risk. As a result, the investor does not require as much of a return as he or she would if the investment was tied up for a longer period. The spread expressed in basis points between long-term and short-term borrowing tends to be greater (on a proportionate basis) when long-term interest rates are higher.

Care should be taken, however, in evaluating interest rate differentials from one financing plan to another. If variable rate financing is selected, then a "worst case/best case" analysis should be prepared to estimate potential savings from a financial plan, as well as the downside risk. In computing financing costs to the issuer, a common technique used for comparing financing plans would be to determine the discounted cash flow requirements of each. Discounted cash flow indicates in present value terms the amount of dollars that would be expected to be paid by the issuer of a certain financing instrument over the amortization period. As a result, the discounted cash flow analysis places alternative financial plans on an "apples versus apples" basis. Noneconomic features of a particular financial plan (increased flexibility, reduced risks, etc.), however, may be significantly more desirable than those of a financial plan which yields a lower discounted cash flow amount.

B. Time Frame and Cost of Issuance

Issuance costs of a financial plan and the related time frame of the plan have an impact on economic feasibility. As a result, these factors should be carefully considered before selecting an appropriate financing plan. Examples of issuance costs include:

- management fees
- legal fees
- closing costs
- remarketing charges
- rating charges
- letter of credit
- printing costs
- trustee costs
- registration/paying agent fees

The utility manager should evaluate what costs are eligible for inclusion in the long-term financing debt vehicle and what costs must be paid at issuance time by the utility.

The financing time frame can be equally important. Conventional long-term financing with revenue or general obligation bonds will likely take at least three to four months from initiation to bond closing. The issuance of tax-exempt commercial paper, on the other hand, may only take four to five weeks. If the utility has an immediate financing need, it may be forced to choose a financing technique to accommodate that need.

C. Risk in Changing Market Conditions

As reflected by the financial marketplace over the last several years, long-term and short-term debt markets can be very volatile. As demonstrated by the stock crash of October 19, 1987, many economic and environmental conditions can trigger extreme reaction by the market. The utility must carefully evaluate potential market conditions that might effect interest rates and other aspects of the financing transaction. If higher interest rates are expected in the future, it may be prudent to use long-term financing. If it appears that interest rates will be decreasing, then a short-term financing instrument may be more appropriate, with conversion to long-term financing when market conditions are more favorable.

D. Potential Tax Law Changes

The current market conditions demonstrate the impact of potential tax law changes. Many investors are very reluctant to invest in currently tax-free debt that may ultimately be ruled taxable. A utility should carefully evaluate potential tax law changes as to how they may affect its interest rates and other features of the debt issued. In some cases, it may be important to accelerate an issue in order to obtain favorable interest rates. On the other hand, certain tax law changes may be advantageous to a utility, and therefore, the utility may desire to wait until the tax laws are changed.

E. The Degree of Public Acceptance

Since a governmental utility is accountable to the public, it is important to evaluate how the public may react to a utility's decision to adopt a specific financial plan. Alternative financial plans have different degrees of risk, cost savings, flexibility, and other features. In some cases, the public may criticize the utility for not selecting a financial plan with attractive interest rates but with significant risks. On the other hand, customers in another geographical area may criticize the utility for accepting a financing plan without considerable economic risk but high interest rates. In order for the utility to make the appropriate trade-offs regarding alternative financial plans, it is necessary to have appropriate input from public officials and other representatives of the customers who will be ultimately affected. The degree of public acceptance can be evaluated by reviewing a community's goals and objectives, public reaction to prior financing decisions, voting history of public officials, and other expressions of public sentiment.

As demonstrated by the above factors, selecting the appropriate financing plan can be a very comprehensive process. Appropriate input from public officials, representatives of the utility customers, and professional advisors is necessary to most appropriately select the optimal course of action. With this input, proper trade-offs can be made considering the economic and noneconomic factors associated with alternative financial plans.

Even with significant input from various interest groups, selecting the appropriate financial plan can be a formidable challenge. As an aid, the utility can use the evaluation matrix presented in Table 6.1. The evaluation criteria discussed above should be included in the matrix and each financial plan should be evaluated as to how it addresses each factor. Numerical values between zero and 10 could be assigned to each cell of the matrix based on how well each criterion is addressed by a particular financial planning scenario. If one evaluation criterion is more important than another, then it should be weighted accordingly. A comparison of the numerical totals of the financial plan would be an indicator of its appropriateness for the utility. The evaluation process should take place over several meetings and a good cross-section of the public sentiment should be represented by participants in the evaluation process. Even though this process is somewhat subjective, it does provide a vehicle for evaluating each of the financing scenarios and developing support for the preferred scenario.

Table 6.1   Financing Plan Selection Evaluation Matrix

Evaluation Criteria	Weight Multiple[a]	Financing Scenarios					
		1		2		3	
		Grade	Total	Grade	Total	Grade	Total
Interest Rate(s) for Financing Plan	1.5	8	12.0	7	10.5	6	9.0
Risk of Changing Market Conditions	1.0	5	5.0	5	5.0	8	8.0
Degree of Public Acceptance	1.5	7	10.5	9	13.5	6	9.0
Impact on Utility Customers	2.0	9	18.0	5	10.0	5	10.0
Time Frame and Cost of Issuance	1.0	5	5.0	7	7.0	8	8.0
Potential Tax Law Changes	1.0	10	10.0	8	8.0	4	4.0
Total Weight			60.5		54.0		48.0
Ranking			1		2		3

[a]Based upon importance of evaluation criteria.

Part II
WATER
and
WASTEWATER
PRICING

7

WATER AND WASTEWATER PRICING PROCESS

7   WATER AND WASTEWATER PRICING PROCESS

With increased environmental and economic attention focused on the water and wastewater industry, the cost of providing water and wastewater services has increased dramatically over the past decade. These increased costs have translated into much higher user charges to utility customers. Water and wastewater user charges, which at one time drew little attention from residential and nonresidential customers, have become a more significant part of the customer's budget. In many communities, rates have doubled or tripled due to several factors influencing pricing: upgraded and more sophisticated treatment systems requiring increased operation and maintenance costs; inflation, which has made a more significant impact on commodities and services affecting utility operations (chemicals, electricity, salaries, etc.); and requirements to staff treatment operations with more qualified managers and operators. As a result of these and other factors, politicians and utility managers have been forced to examine more carefully water and wastewater user charges and pricing affecting their community.

A. Water and Wastewater Utilities as an Enterprise Fund

In a governmental environment, the accounting for water and wastewater operations should be established as an enterprise fund. In the governmental industry accounting book *Governmental Accounting, Auditing, and Financial Reporting* (Chicago: National Committee on Governmental Accounting Publications No. 18, 1968), "enterprise fund" is defined as:

a fund established to account for operations (a) that are financed and operated in a manner similar to private business enterprises — where the intent of the governing body is that the costs (expenses, including depreciation) of providing goods or services to the general public on a continuing basis be financed or recovered primarily through user charges; or (b) where the governing body has decided that periodic determination of revenues earned, expenses incurred, and/or net income is appropriate for capital maintenance, public policy, management control, accountability, or other purposes. Examples of Enterprise Funds are those for water, gas, and electric utilities; swimming pools; airports; parking garages; and transit systems.

As an enterprise fund, water and wastewater operations are viewed as a business. As a result, appropriate business principles related to cost identification, cost-effectiveness, and financial reporting should be addressed by the enterprise organization.

As indicated by the definition of an enterprise fund, appropriate fees and charges should be established to ensure that the organizations can operate on a self-sustaining basis. In a water and wastewater utility, the majority of revenue is normally derived through user charges. User charges are defined as: *fees, rates, assessments, and billings that are charged to beneficiaries of water and wastewater services.* User charges are a way of recovering cost for providing a service from those that benefit directly from that service. The total process of (1) identifying water and wastewater costs, (2) allocating costs to utility beneficiaries, and (3) designing rate structures to recover allocated costs is defined as a *user charge system.* Another common term for a water and wastewater user charge system is water and wastewater pricing. The purpose of this part of the book is to discuss the water and wastewater pricing process. More specifically, we will discuss in detail how water and wastewater costs are identified and how rates are developed. In addition, we will discuss the differences in water and rate structures for utilities around the country.

B. Characteristics of an Effective Pricing Structure

In establishing user charges, utility managers and political leaders have to address two major issues:

- Which costs should be recovered through user charges?

- How should a pricing structure be designed to ensure that utility and community objectives are being appropriately addressed?

The utility must make certain that adequate revenues are recovered through user charges, allowing the utility to operate on a self-sustaining basis. Adequate revenues ensure that (1) salaries can be established at a level to attract qualified managers, supervisors, clerks, and operators; (2) necessary chemicals and supplies are readily available for operational needs; (3) electric and gas bills can be paid in a timely fashion; (4) capital replacements can be made in an effective manner; and (5) other capital needs can be addressed appropriately. On the other hand, setting user charges too high unfairly taxes users and may encourage utilities to be less fiscally responsible in establishing operating and capital programs.

The second major goal of an effective water and wastewater pricing structure is to establish a structure which attempts to maximize the goals and objectives of the utility and the community. In other words, a pricing structure should be designed to promote the planning philosophies of a service area, assuming that these philosophies do not conflict with basic pricing principles. Factors that should be considered in designing appropriate rate structures include:

- *Equity or fairness:* Costs should be recovered from customers or customer classes in proportion to the costs of providing service to these beneficiaries.

- *Legality:* A user charge structure should comply with appropriate local, state, and federal legal requirements such as the EPA user charge regulations.

- *Impact on customers:* The economic impact that a proposed user charge structure will have on water and wastewater customers should be considered. When a utility modifies its rate structure, costs are typically redistributed among customer classes and significant impacts can affect certain customer classes.

- *Avoidance of discriminatory relationships:* The user charge system should avoid placing an unfair cost recovery burden on any particular class of customer. In some states, discounted rates (such as lifeline) and subsidies have been considered discriminatory practices.

- *Simplicity:* The user charge structure should be easy to understand and to discuss with public officials and customers.

- *Implementation:* The rate structure should be easy to implement, administer, and update.

- *Competitiveness:* Rates and rate structures of similar and adjacent communities should be considered.

- *Conservation:* Efficient use of water and wastewater resources should be encouraged.

It is important to note that several of these pricing criteria can be conflicting. For example, establishing a rate structure with increasingly priced rate blocks may directly address the conservation objective but may be contrary to the cost-of-service and financial sufficiency objectives. Establishing a cost-of-service user charge structure with numerous user classes may be considered to be equitable but will likely be complex and possibly be very difficult to implement and maintain. These trade-offs show that rate making is an art that carefully considers alternative rate-setting techniques to optimize a community's pricing objectives.

C. Approach to Establishing User Charges

Pricing structures can vary significantly—from being very simple to very complex. The pricing process for various user charge structures, however, is similar in that it involves a three-step process. As depicted in Figure 7.1, this process includes:

- Step 1: Identify revenue requirements.
- Step 2: Determine cost of service.
- Step 3: Design rate structure.

STEP 1: IDENTIFY REVENUE REQUIREMENTS

The first step in any rate-setting process is to identify those costs that must be recovered through water and wastewater user charges. In government utilities, these costs typically consist of operating and capital costs. Operating costs are those required to operate and maintain the utility on an ongoing basis. Examples of these costs would include salaries, electricity, chemicals, and other recurring costs. For a government utility establishing its revenue requirements on a cash basis, capital costs typically include debt service; capital costs financed through current revenues; and contributions to special replacement, improvement, and expansion funds. In many rate-setting processes it is important to project operating and capital costs over an extended period so that fluctuations in potential rates can be evaluated. It is essential that revenue requirements be sufficient to provide for adequate facilities, to allow for proper replacement and maintenance, and to ensure that the utility is operated on a self-sustaining basis.

STEP 2: DETERMINE COST OF SERVICE

After revenue requirements have been identified, the next step is to allocate these requirements to classes of utility customers based on the cost to serve these customers. Classes of customers might include:

- residential (inside city, outside city, single-family, multi-family, etc.)
- commercial
- industrial
- institutional (hospitals, schools, colleges, etc.)
- governmental (water districts, municipalities, military bases, etc.)

One goal of an effective rate-setting process is to choose appropriate allocation factors which allocate costs to customer classes based upon the cost of providing services to that class of customer. Examples of these cost factors would include hourly, daily, and average customer demand; customer location within the system; facilities constructed to serve a specific customer or customer class; wastewater strength; and other variables.

STEP 3: DESIGN RATE STRUCTURE

After costs have been allocated to appropriate user classes, it is then necessary to design a rate structure for appropriately charging customers. In some cases, special rate structures are designed for each class of customer. In other situations, a combined rate structure is

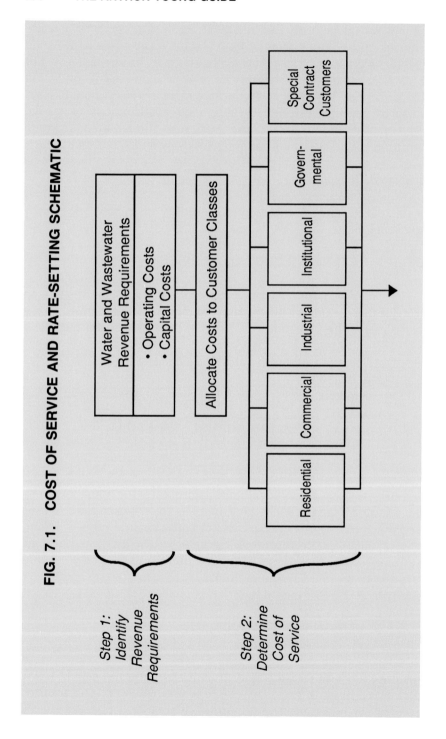

FIG. 7.1.   COST OF SERVICE AND RATE-SETTING SCHEMATIC

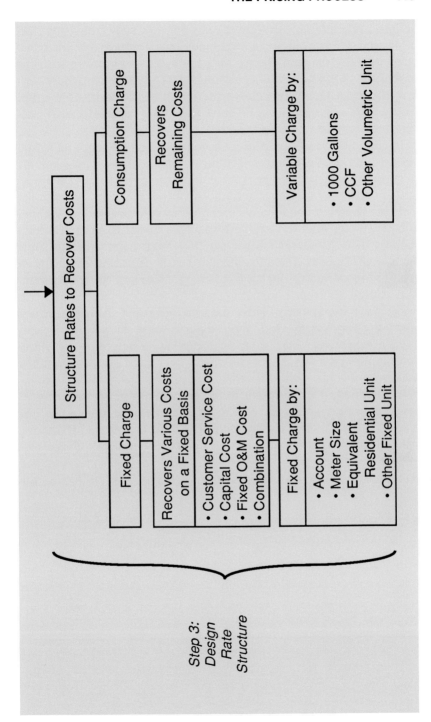

designed that attempts to recover costs appropriately from each class of customer. For example, a multiblock rate structure is often developed whereby the unit charge for each block generally reflects the cost of providing service to a particular class of customer. Under such a structure, each customer class typically has the majority of its usage falling within that block. In Figure 7.1, we have depicted a rate structure with only one consumption block.

In designing a rate structure, it is common that a water and wastewater utility establish both a fixed or minimum charge and a volume or consumption charge. The fixed charge is designed to recover a base amount from the customer periodically, often monthly or quarterly. This base amount could include service costs (meter reading, billing, collection, customer service, etc.); some portion of fixed capital costs such as debt service and other capital-related costs; and, in some cases, fixed operating and maintenance costs. In flat-rate systems all costs are recovered through a fixed charge. The volume or consumption charge recovers remaining costs from customers based upon usage. In most systems, water usage is measured by water meters which are periodically read. Wastewater volume charges may be calculated based on actual water consumption, a percentage of water consumption, or actual measurement by special wastewater flow meters.

In summary, it is important to recognize that the rate-setting process should be dynamic and structured to address the pricing philosophies of the utility and the community. Chapters 8, 9, and 10 discuss the cost determination and rate-setting process in detail. Chapter 11 then presents the results of a survey of water and wastewater rates and rate structures among the 120 largest government utilities in the United States.

8 IDENTIFICATION OF REVENUE REQUIREMENTS

8   IDENTIFICATION OF REVENUE REQUIREMENTS

As depicted in Figure 7.1 in Chapter 7, identifying the appropriate amount of revenue requirements is an important first step in establishing user charges. Recovering appropriate revenue requirements ensures financial sufficiency for the utility.

Revenue requirements are usually divided into two major categories: (1) operating and maintenance costs and (2) capital costs. O&M costs are usually routine or periodic costs incurred by a water and wastewater utility in providing service on an ongoing basis. Capital costs relate to capital items such as equipment or facilities that provide benefits for greater than one year. In some cases, return on investment as discussed below is included as a part of revenue requirements.

A. Approaches to Establishing Revenue Requirements (O&M and Capital Costs)

In establishing revenue requirements, a utility should follow either the utility approach or the cash-needs approach. The use of these approaches has been the subject of contention by many rate technicians. These approaches are discussed below.

1. UTILITY APPROACH

The utility approach to determining revenue requirements is followed by almost all investor-owned and governmental utilities which are regulated by state public service agencies. The utility approach allows the utility to recover operating and capital cost as determined by generally accepted accounting principles. In addition, the utility is allowed to earn a return on its investment in plant in-service and other capital facilities.

A historical test year is first established (typically for the most recent accounting period), and operating expenses and capital costs are calculated for this test year. O&M costs can be projected for the rate recovery period by estimating inflation, additions to staff, salary adjustments negotiated under union contracts, adjustments for chemicals and supply costs, and other pertinent factors. Capital costs can be estimated based upon projected depreciation and adjusted for additions, retirements, contributions-in-aid-of-construction, and customer capital advances. Certain public service regulatory agencies, however, are very restrictive in what futuristic costs they will allow a utility to recover. For example, inflationary costs may be disallowed, since they are not known and measurable. Many regulatory bodies, however, will usually permit future costs projections that are known and measurable in setting rates.

Under the utility approach, a return is calculated on the allowable investment by the owner of the utility. The utility's investment is defined as a "rate base," and the return represents the earnings to the owners or stockholders. The investor's return should provide for the recovery of certain inflation costs (not included in operating costs), the retirement of principal on debt service, the funding of certain capital items, and/or a payback (dividend) to the owners of the utility. If the utility is governmental, the return is still appropriate, although the utility is "nonprofit." As with investor-owned utilities, the return is used to retire principal on debt and for funding of certain capital items. The "dividend" for government utilities may be eliminated, however, since a return or "profit" component may be excluded from revenue requirements.

The major advantage of the utility approach is that there is typically less interpretation when establishing revenue requirements than under the cash-needs approach. In other words, the utility approach provides for a less subjective methodology for identifying revenue requirements. A major disadvantage of the utility approach is that in a governmental environment, revenue requirements that would be recovered under the utility approach, even though levelized, could be significantly more or less than is required for cash-flow purposes.

2. CASH-NEEDS APPROACH

Under the cash-needs approach, user charges are structured to recover specific cash requirements for O&M and capital. In other words, the cash budget requirements are used as the basis for establishing user charges.

The major difference between the utility approach and the cash-needs approach is the way in which capital costs are included as a part of revenue requirements. Under the utility approach, depreciation, interest on debt service, and return on rate base provide the basis for capital requirements. Under the cash-needs approach, several capital requirements are typically included: debt service (principal and interest), capital outlay (pay-as-you-go capital), and contributions to various reserve funds. Major capital outlays would usually be handled by new bond issues or the use of reserve funds.

A major advantage of the cash-needs approach is that it provides great flexibility to a utility in that it establishes rates to recover the cash requirements of the utility. In other words, there is more latitude when the utility schedules capital expenditures as a component of revenue requirements. For this reason, the cash-needs approach is preferred by most government utilities. In Table 8.1, we summarize the advantages and disadvantages of the cash-needs and utility approaches.

Table 8.2 demonstrates how revenues under the two approaches may vary. Under the cash-needs approach, revenue requirements are established at $27,000,000. On the other hand, revenue requirements under the utility approach are established at $25,000,000. In this example, the major difference in the requirements results primarily from significant reserve fund contributions required under the cash-needs approach. Capital funding approaches vary extensively from utility to utility, however, and these approaches directly affect how capital costs are included as revenue requirements under the cash-needs and utility approaches.

Most governmental utilities adopt the cash-needs approach in identifying their revenue requirements. For the remainder of the chapter, we will assume that the utility follows the cash-needs approach for establishing user charges. Assuming the cash-needs approach, we will discuss in detail the operating costs and capital costs that would normally be identified as part of the revenue requirements.

B. Operating and Maintenance Costs

For water and wastewater utilities, O&M costs are defined as costs that are ongoing and recurring, and that are generally incurred during a utility's accounting period. Examples of major categories of O&M

Table 8.1   Comparison of Cash-Needs Approach vs Utility Approach

ADVANTAGES

Utility Approach	Cash-Needs Approach
• Is less subjective. • Better matches cost of service with beneficiary use (e.g., used and useful analysis). • Is more consistent with generally accepted accounting principles.	• Is consistent with governmental budget practices. • Can be easier to understand because it matches revenue with cash needs. • Is consistent with bond rating agencies' evaluation of revenue-producing capability. • Provides increased flexibility. • Bond covenants are predicated on cash needs. • Is generally accepted by governmental utility industry.

DISADVANTAGES

Utility Approach	Cash-Needs Approach
• May generate insufficient or excessive revenue for cash needs. • Is not generally accepted in governmental water and wastewater utility industry unless the utility is regulated. • Provides less flexibility. • Is more difficult to explain to customers/policy-makers.	• Could result in large net profits or losses if financial statements are prepared in accordance with generally-accepted accounting principles. • Can be more difficult to match the recovering capital costs in varying periods. • Is not usually accepted as a valid method by state public service commissions.

costs would be salaries and wages, purchased power, chemicals, materials and supplies, and rental of equipment.

O&M costs should not relate to the cost of capital facilities. In addition, O&M costs should be differentiated from depreciation. Unlike O&M costs, depreciation relates strictly to capital items and attempts to allocate the cost of capital items over a series of accounting periods greater than one year. Under the utility approach, depreciation is included in the revenue requirements. Under the cash-needs approach, it is not included.

1. CLASSIFYING O&M COSTS

In order to properly account for O&M costs, it is necessary to develop a common yardstick for classifying costs consistently from year to year. Specifically, O&M costs should be classified in a manner to:

Table 8.2   Revenue Requirements Under Cash-Needs and Utility Approaches

Revenue Requirements Item	Cash-Needs	Utility
Operating Costs	$10,000,000	$10,000,000
Depreciation[a]		2,000,000
Return on Investment[b]		6,000,000
Debt Service[c]		
• Principal	4,000,000	
• Interest	7,000,000	7,000,000
Minor Capital Outlay	1,500,000	
Reserve Fund Contribution		
• Operating	500,000	
• Replacement	1,000,000	
• Expansion	1,000,000	
• Insurance	500,000	
• Rate Stabilization	500,000	
• Debt Service	1,000,000	
Total	$27,000,000	$25,000,000

[a]On an investment of $100,000,000 (acquisition amount), with no contributions-in-aid-of-construction, and using a 2% composite depreciation rate.
[b]Rate of return is established at 6% (weighted cost of debt).
[c]Assumes $100,000,000 was bonded at 10% interest and amortized over 25 years with principal of $4,000,000 and interest of $7,000,000 during the rate recovery projection period.

- Support cost-of-service and rate-making calculations.
- Provide proper monitoring and reporting of each O&M cost.
- Enhance comparability of costs among water and wastewater utilities.
- Provide appropriate information to utility managers for operating the utility in a cost-effective manner.

The most effective means of classifying and tracking O&M costs is through an effective chart of accounts. A chart of accounts is a means of classifying all assets, liabilities, costs, revenues, and other accounting transactions on a consistent basis. As a utility completes a financial transaction, a record of that transaction is tracked into the appropriate account within the chart of accounts. In other words, for an O&M cost to get appropriately classified, the chart of accounts is used to properly "code" the O&M cost item.

Charts of accounts have been recommended for water and wastewater utilities by the National Association of Regulatory Utility Commissioners (NARUC) and the National Council of Governmental Accounting (NCGA). For a water utility, a NARUC chart of accounts has been developed for different sizes of water utilities — Classes A, B, and C. A similar classification system has been developed for wastewater utilities. Typically, the larger and more complex the utility, the greater the need is for a more detailed chart of accounts. A detailed chart of accounts with account descriptions for water and wastewater utilities can be obtained by writing NARUC in Washington. These charts of accounts provide the industry standard for tracking O&M costs.

Table 8.3 presents the account titles for the NARUC chart of accounts for O&M costs for Class A water utilities. This chart of accounts was developed by first identifying categories of O&M costs, called "objects." These objects reflect characteristic costs that are incurred commonly in many aspects of water utility operations and administration. For example, salaries and wages are common O&M cost objects for each functional section within the organization. As presented in Table 8.3, examples of O&M objects would include "Salaries & Wages," "Purchased Water," "Chemicals," "Contractual Services," and "Materials and Supplies."

The next step in developing the chart of accounts is to identify logical, functional categories to which these O&M cost categories relate. As presented in Table 8.3, these functional categories include:

- source of supply
  - operations
  - maintenance
- water treatment
  - operations
  - maintenance
- transmission and distribution
  - operations
  - maintenance
- customer accounts
- administrative and general

The O&M chart of accounts matrix is then finalized by identifying which O&M cost objects are relevant to each functional category. For example, all O&M objects except (1) "Advertising Expense," (2) "Regulatory Commission Costs — Amortization of Rate-Case Expense," and (3) "Regulatory Commission Cost — Other" are relevant to the functional category "Source of Supply." The completed chart of

Table 8.3   Water Operation and Maintenance Expense Accounts for Class-A Utilities[a]

Account #	Object Descriptions	Functional Expense Categories							
		Source of Supply: Operations	Source of Supply: Maintenance	Water Treatment: Operations	Water Treatment: Maintenance	Trans. & Dist. Operations	Trans. & Dist. Maintenance	Customer Account	Admin. & General
601	Salaries and Wages—Employees	601.1	601.2	601.3	601.4	601.5	601.6	601.7	601.8
603	Salaries and Wages—Officers, Directors, and Majority Stockholders	603.1	603.2	603.3	603.4	603.5	603.6	603.7	603.8
604	Employee Pensions and Benefits	604.1	604.2	604.3	604.4	604.5	604.6	604.7	604.8
610	Purchased Water	610.1	–	–	–	–	–	–	–
615	Purchased Power	615.1	–	615.3	–	615.5	–	615.7	616.8
616	Fuel for Power Production	616.1	–	616.3	–	616.5	–	616.7	616.8
618	Chemicals	618.1	618.2	618.3	618.4	618.5	618.6	–	–
620	Materials and Supplies	620.1	620.2	620.3	620.4	620.5	620.6	620.7	620.8
631	Contractual Services—Eng.	631.1	631.2	631.3	631.4	631.5	631.6	631.7	631.8
632	Contractual Services—Acct.	632.1	632.2	632.3	632.4	632.5	632.6	632.7	632.8
633	Contractual Services—Legal	633.1	633.2	633.3	633.4	633.5	633.6	633.7	633.8
634	Contractual Services—Management Fees	634.1	634.2	634.3	634.4	634.5	634.6	634.7	634.8
635	Contractual Services—Other	635.1	635.2	635.3	635.4	635.5	635.6	635.7	635.8
641	Rental of Building/Real Property	641.1	641.2	641.3	641.4	641.5	641.6	641.7	641.8
642	Rental of Equipment	642.1	642.2	642.3	642.4	642.5	642.6	642.7	642.8
650	Transporation Expenses	650.1	650.2	650.3	650.4	650.5	650.6	650.7	650.8
656	Insurance—Vehicle	656.1	656.2	656.3	656.4	656.5	656.6	656.7	656.8
657	Insurance—General Liability	657.1	657.2	657.3	657.4	657.5	657.6	657.7	657.8
658	Insurance—Workman's Comp.	658.1	658.2	658.3	658.4	658.5	658.6	658.7	658.8
659	Insurance—Other	659.1	659.2	659.3	659.4	659.5	659.6	659.7	659.8
660	Advertising Expense	–	–	–	–	–	–	–	660.8
666	Regulatory Commission Expenses—Amortization of Rate Case Exp.	–	–	–	–	–	–	–	–
667	Regulatory Commission Expenses—Other	667.1	667.2	667.3	667.4	667.5	667.6	667.7	667.8
670	Bad Debt Expense	–	–	–	–	–	–	670.7	–
675	Miscellaneous Expenses	675.1	675.2	675.3	675.4	675.5	675.6	675.7	675.8

[a]From the National Association of Regulatory Utility Commissioners (NARUC) and the National Council of Governmental Accounting (NCGA).

accounts is then used to appropriately code all water utility O&M costs. A similar chart of accounts would be developed for wastewater accounts.

A major benefit of this comprehensive breakdown is that it classifies O&M costs in such a manner that will allow for efficient calculation of cost of service to customer classes. In addition, the NARUC chart of accounts is structured so that it can usually be adapted to the size and accounting sophistication of different utilities. For example, large utilities with comprehensive accounting and work order systems might track O&M costs by using the chart of accounts for Class A utilities. Smaller utilities with less precise accounting systems might use the chart of accounts for Class C utilities.

2. O&M COST CONSIDERATIONS IN GOVERNMENTAL WATER UTILITIES

One drawback of the NARUC expense account system is that the structure may not relate directly to the organizational and budget structure of the water or wastewater utility. For example, a government utility may be organized as depicted in Figure 8.1. In addition, the accounting system and related chart of accounts many times are structured by organizational unit. In the utility depicted in Figure 8.1, costs might be tracked by the organizational section—"Demand Forecasting" (in the engineering department)—rather than by an appropriate NARUC chart of accounts.

In addition, a government utility usually must adopt the chart of accounts of its local government accounting system. (Many times, this is mandated by state law.) The state identifies specific "object" expense accounts that must be applied across the organizational units of the local government. A common grouping for O&M cost objects in a government environment follows:

- personal services
- contractual services
- commodities
- administration and indirect costs

Under each category there could be numerous expense objects, some of which could be similar to the NARUC expense objects, while others could be very different. In addition, many of the objects might relate strictly to other city or county departments and have no relevance to the water or wastewater departments. In such cases, object

FIG. 8.1.   ORGANIZATION OF A GOVERNMENT UTILITY

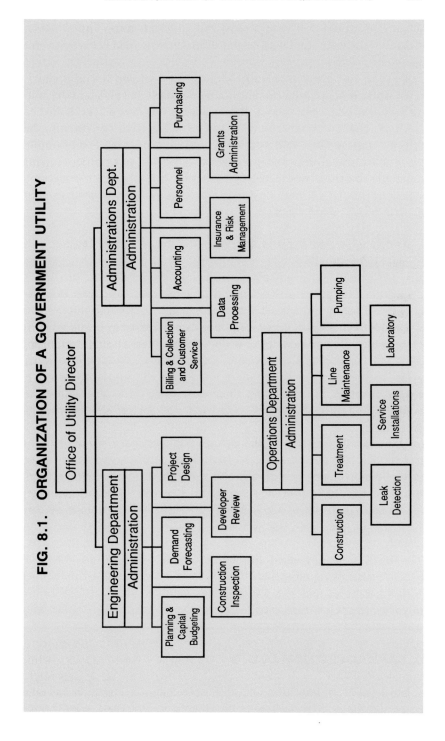

categories would be used across all governmental departments, including the water and wastewater departments. This process is similar to the NARUC process whereby NARUC objects are used across functional cost components of the water and wastewater utilities.

To provide adequate information for multiple purposes, many utilities adopt two cost accounting systems: one consistent with the NARUC chart of accounts, to provide for effective rate setting and utility comparability, and another to provide for cost accountability by organizational unit and for consistency with accounting systems mandated by state and local governments.

3. INDIRECT O&M COSTS

As discussed above, many governmental water and wastewater utilities are departments of city or county governments. In such cases, the city and county governments provide support services to the utility department. Support services might include such functions as planning, purchasing, personnel, accounting, and data processing.

To identify all O&M costs, it is important to identify indirect O&M costs incurred on behalf of the water and wastewater departments by other city or county departments. Otherwise, all costs of providing water and wastewater services are not identified, and cost comparisons with other utilities using these costs directly are not valid. In addition, the utility is not recovering the true cost of service from its customers.

To address this issue, many local governments have developed central service cost allocation plans to allocate indirect support costs to service departments. In such cases, it would be a simple matter to take the allocation of indirect costs to the water and wastewater departments and identify it as an O&M cost in the revenue requirements. If such a plan does not exist, water and wastewater indirect costs can usually be estimated by using some logical basis for each indirect cost. The basis normally used has some logical relationship with the expense item being allocated. For example, the number of personnel within the water and wastewater departments to total city personnel might be the percentage used to allocate personnel costs to water and wastewater revenue requirements.

4. CAPITALIZED O&M COSTS

Most water and wastewater utilities are capital-intensive. In some cases, capital-related revenue requirements (debt service, pay-as-you-go capital, reserve fund contributions, etc.) can be as high as 50–75% of annual revenue requirements.

From a revenue requirements perspective, it is important to recognize the significance of capitalizing O&M costs. When O&M costs are capitalized, they are not normally included as a part of revenue requirements to be recovered through rates. Instead, they are typically recovered through a capital funding source. In our example, the portion of engineering costs related to the wastewater transmission line might be capitalized and recovered through the funding source for the transmission line. Funding sources could include bond proceeds, capital expansion funds, capital recovery funds, and renewal and replacement funds. In utilities where significant construction is performed by in-house personnel, much of the operating budget is capitalized and recovered through typical capital sources.

5. CAPITAL COSTS TREATED LIKE O&M COSTS

In some instances, operating budgets of governmental utilities include certain capital items to be funded through current revenues. These capital items are normally referred to as "pay-as-you-go" capital. Good examples of pay-as-you-go capital items would include vehicles, motors, pumps, water meters, and other high-use items with lives usually less than 8–10 years. Even though these items are clearly capital items and benefit more than one accounting period, they get included in the annual operating budget and are dealt with for revenue requirements in a similar manner as O&M expenses.

Capital items that are funded through operating budgets are still capital items and are capitalized for accounting purposes. In other words, they are set up as an asset and are depreciated over a period of years. If rates are based upon a cash-needs approach, these capital items are financed through rates. If the utility approach is adopted, these capital items are funded through depreciation. In some cases, when capital expenditures of these items are relatively constant over a period of years, then cash expenditures for pay-as-you-go capital will approximate what depreciation would be on the same items.

6. NON-ANNUAL O&M COSTS

Some O&M costs have neither the characteristics of an annual O&M cost nor those of a cost that gets capitalized. A good example of a non-annual O&M cost would be the painting cost associated with water storage tanks. This expense does not create a new asset but provides maintenance to an existing asset. This expense has the characteristics of an O&M cost but might be incurred only once every 8–10 years.

Many investor-owned utilities capitalize non-annual O&M costs as a "deferred debit" during the year in which the cost was incurred. The deferred debit is amortized over the period to which the cost relates, very similarly to a fixed asset or a prepaid expense. Other non-annual O&M costs that might get capitalized include costs for rate hearings, planning studies, and certain maintenance activities. Unamortized costs are normally allowed as a component of the rate base on which the investor-owned utility is allowed to earn a return.

In some instances, non-annual O&M costs for investor-owned utilities should be expensed annually. These instances would occur when the costs are spread out in such a way that approximately the same amount of cost is incurred annually. For example, a water utility may have 10 elevated storage tanks, with one tank being painted annually. By the time the 10th tank is painted in the 10th year, it is time to repaint the first tank, and so on. In such a case, the cost of tank painting would be expensed annually and included as a new revenue requirement each year.

For government utilities not regulated by public service commissions and developing their revenue requirements on a cash-needs basis, more flexibility is allowed in including non-annual O&M costs in revenue requirements. In governmental utility environments, non-annual O&M costs could be recovered through the capital budgets or as an O&M cost, or they could be amortized over a period of years. For purposes of revenue requirements, the circumstances surrounding the O&M cost should dictate how it is handled.

7. ESTIMATING O&M COSTS FOR FUTURE YEARS

Many times, revenue requirements are based on future rate years. In some instances, multiyear rates would require several future rate years to be considered. This practice is particularly true for governmental utilities not governed by public service commissions.

When a future rate year is used, it is common practice to use the proposed operating budget of an unregulated government utility as a basis for estimating next year's O&M costs as a part of revenue requirements. When a proposed operating budget is not available, or when future years beyond the budget year are required, it is necessary to escalate O&M costs for these future periods. In such instances, it is necessary to consider (1) inflation, and (2) the impact related to increased system demand. In other instances, seasoned judgment is important when this judgment can be supported by reasonable assumptions.

Inflation

To estimate the impact of inflation on O&M costs, it is common to use inflationary indexes or other relevant bases. In some cases, future O&M costs are established by contract. For example, the utility may have a contract for a service over a five-year period, with a known amount identified. In other instances, the impact of inflation on costs should be estimated. In such instances, escalation indexes such as Engineering News-Record (ENR) indexes, Municipal Cost Index, and EPA indexes could be considered in estimating the impact of inflation on O&M costs. It is generally preferable that most major categories of costs (salaries, utilities, commodities, etc.) be evaluated individually and an appropriate index used which best relates to the category of expense. In some cases, such as with governmental salaries, historical trends in inflation may be appropriate for establishing future salary amounts.

Impact Related to Increased System Demand

In addition to inflation, it is necessary to consider the impact of increased demand on O&M costs. As demand increases, more utility costs and chemicals are typically required in order to provide for this demand. In the short term, some costs may not be affected by increased demand. For example, staffing at a water treatment plant may remain the same until a new plant is built. In such instances, no increase to these costs should be considered related to increased demand.

8. SUMMARY

In summary, O&M costs are a major component of water revenue requirements and it is important to classify O&M costs through an appropriate chart of accounts. For cost-of-service purposes, a NARUC chart of accounts provides the most appropriate O&M classification system. For identifying O&M costs, indirect costs should be considered. It is also important to recognize that some O&M costs are capitalized, some capital costs are treated like O&M costs, and special treatment is used for certain non-annual O&M costs. In addition, it is

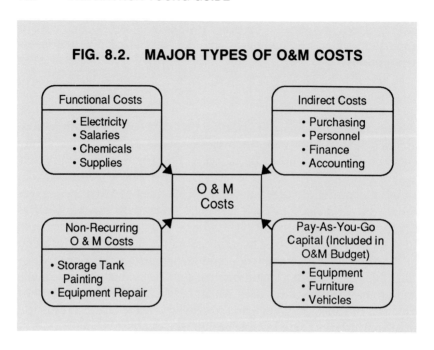

FIG. 8.2.   MAJOR TYPES OF O&M COSTS

accepted practice to use appropriate inflation indexes in escalating O&M costs. Major types of O&M costs are depicted in Figure 8.2.

C. Capital Costs

Under the cash-needs approach, it is important to identify the cash that is needed from user charges to support capital programs and related capital expenditures. In Part I, we discussed how capital items get financed and related costs recovered. Capital cost recovery methods include bonds, assessments, grants, impact fees, and user charges. An example was provided of what capital costs are typically recovered through user charges. Specifically, these capital costs include:

> • *Debt service:* Debt service consists of principal and interest that must be repaid on bonds issued by the utility. The amount of principal and interest on bonds will vary by the interest rate, amount of the issue, reserve fund requirements, amortization period, coverage require-

ments, and other financial considerations. A detailed discussion of bonds is presented in Chapter 4.

- *Revenue financed capital outlays:* Many utilities choose to fund some capital items through user charges. These items are typically defined as pay-as-you-go capital items, as discussed above, that would tend to be highly used and typically last anywhere from 3 to 10 years. Examples of these items might include vehicles, motors, pumps, furniture, etc. In some cases, these capital outlays might be significant and could relate to major renewal, replacement, and improvement facilities or projects. The level of expenditures for these items normally varies significantly from year to year.

- *Reserve fund contributions:* Bond ordinances and good fiscal management and rate-setting practices dictate that certain reserve funds be established. Several cash reserve funds that could be established include:
  - Operating fund: This fund is typically established to pay for operating needs and provides the utility with a working capital reserve that would be necessary for cash flow purposes.
  - Capital replacement fund: This fund is established to replace various utility assets as appropriate. Included in this category would normally be replacement items similar to minor capital outlay items discussed above. In addition, some utilities choose to include in their replacement fund monies for replacing major facilities of the utility, such as distribution lines, collection lines, storage facilities, etc.
  - Capital expansion fund: Many utilities create a fund that can be used for expanding the utility to promote growth. This fund is typically supported by capital recovery charges and customer contributions.
  - Insurance reserve fund: This reserve is established in order to provide self-insurance. Rather than paying insurance premiums, or where insurance is not available, the utility might develop a self-insurance fund to protect themselves against uncertainties and catastrophic events.
  - Debt service reserve fund: This fund is typically a requirement of a bond ordinance and would provide

additional protection to the bond holder. The debt service reserve could be used if the utility was unable to pay debt service from current revenues. The debt service reserve fund might be bonded or funded through current revenues.

–Rate stabilization fund: Revenue requirements and usage often vary significantly from year to year. In such cases, the utility may wish to stabilize rates rather than have major rate adjustments from year to year. As a result, a rate stabilization fund can be established to cover revenue requirements in years where rates may be insufficient to generate appropriate revenues.

In projecting capital costs over extended recovery periods (three to five years), capital costs must usually be inflated. Popular escalation indexes would include the Handy Whitman Index, the ENR indexes based on the type of facility involved, and construction price indexes as developed by the Environmental Protection Agency. If a good history of capital cost increases is available locally, these local statistics sometimes provide a better basis for escalating capital costs.

9   DETERMINATION OF COST OF SERVICE

9 DETERMINATION OF COST OF SERVICE

After revenue requirements have been identified, the second step in the pricing process is to determine the cost of providing service to classes of water and wastewater customers. As depicted in Figure 9.1, cost of service is determined by reducing revenue requirements developed in Step 1 of Figure 7.1 by revenue offsets, and then allocating net revenue requirements to customer classes by some indicator(s) of how much of net revenue requirements should be allocated to each class. This chapter discusses the issues involved in the allocation process.

A. Revenue Offsets to Revenue Requirements

Almost every utility generates some revenue from sources other than user charges. As depicted in Figure 9.2, revenues can be derived from operational or non-operational sources. Operational revenues are typically generated from specific or miscellaneous services provided by the utility. Non-operational revenues typically relate to investment income; special capital charges such as assessments; impact fees; franchise fees; or nonrecurring sales of materials, facilities, and other assets. Specific service revenue and non-operational revenues will be discussed below.

1. REVENUES FROM SPECIFIC SERVICES PROVIDED BY THE UTILITY

Specific services are services that (1) are secondary to the primary purpose of the utility to provide general water and wastewater service, and (2) provide direct benefit to a particular customer or class of customers. Specific service revenues are typically removed from reve-

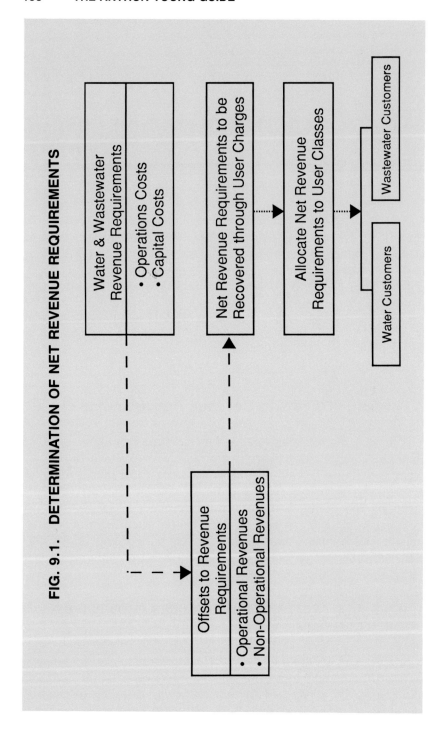

FIG. 9.1.   DETERMINATION OF NET REVENUE REQUIREMENTS

Water & Wastewater Revenue Requirements
• Operations Costs
• Capital Costs

Offsets to Revenue Requirements
• Operational Revenues
• Non-Operational Revenues

Net Revenue Requirements to be Recovered through User Charges

Allocate Net Revenue Requirements to User Classes

Water Customers

Wastewater Customers

FIG. 9.2. IDENTIFYING OPERATIONAL AND NON-OPERATIONAL REVENUES

Offsets to Revenue Requirements

- Operational Revenues
- Non-Operational Revenues

Operational Revenues - Specific Service Charges	Non-Operational Revenues

• Water and Sewer Connections
 	
• Delinquent Disconnections and Reconnections
 	
• Service Installations
 	
• Engineering Services
  - Design Review
  - Inspection
 	
• Meter Testing
	• Interest Income
	
	• Special Assessments
	
	• Impact Fees
	
	• Sale of Assets
	
	• Franchise Fees

Design Specific Service Charges	Determine Use of Non-Operating Revenues to Offset Revenue Requirements
Revenue from Operational Sources	Revenue from Non-Operational Sources

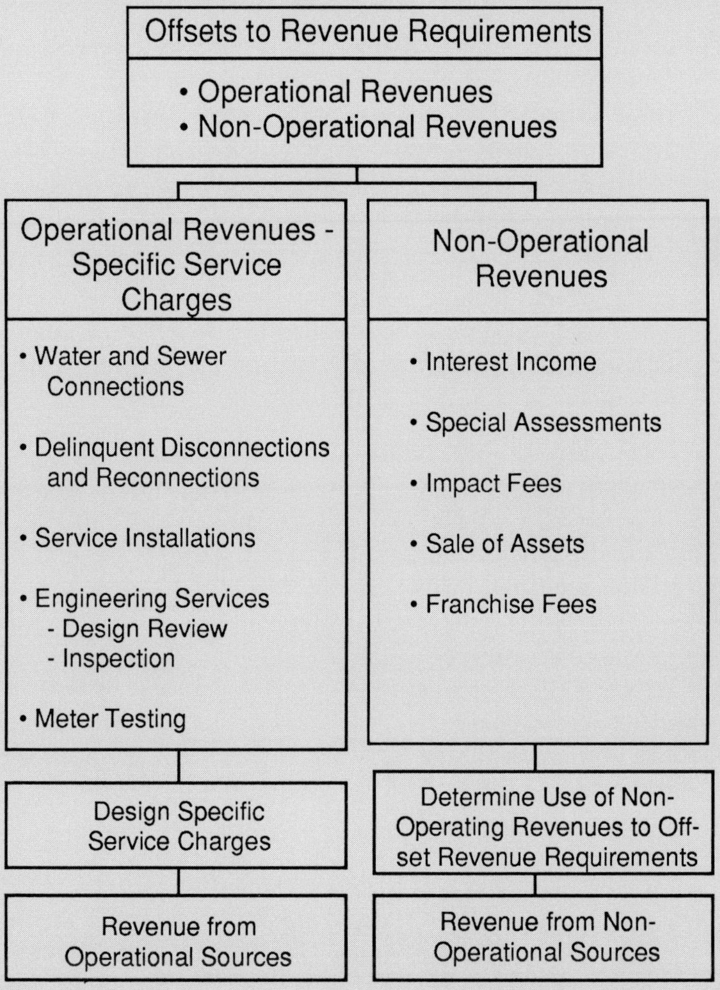

nue requirements in that specific service costs should be recovered from the direct beneficiary of the service rather than from general water and wastewater customers.

Examples of specific services include:

- water service lines and sewer lateral connections — connecting a new customer from his or her property line to the water or sewer main in the street.

- service installation — establishing service to a new customer who moves into a premises that is already physically connected to the system. Typically involved in a service installation would be a water meter turn-on as well as administrative processing to set up a new account.

- engineering services — services typically provided by the utility's engineering department to developers. This assistance would include reviewing designs of water and wastewater systems to be installed by the developer, and ultimately inspecting the construction performed by the developer.

- septic disposal and treatment — providing special treatment services to haulers of septic waste.

Other specific service charges that are charged by water and wastewater utilities are listed below:

- new account charge
- returned check charge
- meter reread charge
- fire protection charge
- meter testing fee
- water sales — fire hydrants
- delinquent charge
- sales of plans and specifications/blueprints
- notification charge
- charge for recovery of damages
- meter turn-off/on charge
- meter removal charge
- water tap tests
- turn-off/on at main charge
- sales of materials

- illegal water use penalty
- sales of equipment
- emergency water turn off/on charge

Certain utilities choose to recover specific service costs through water and wastewater rates rather than through specific service charges. In determining whether a specific service charge should be established, certain key issues should be addressed:

- Does the specific service occur with sufficient frequency to warrant a special charge being developed?

- Is good work order data or reasonable estimates available for establishing the charge?

- Are the revenues sufficient to justify the cost associated with collecting the revenues?

If the answer to any of these questions is no, then it is likely that establishing a specific service charge for the service under consideration would be inappropriate. In addition, when implementing a new specific service charge or increasing an existing charge significantly, the impact on demand should be considered carefully in projecting revenues.

2. REVENUE FROM NON-OPERATING SOURCES

Utilities also generate revenues from non-operating sources. These revenues are typically derived either through investing surplus funds, sale of assets, or special capital-related charges to customers.

Interest income is normally earned on excess operating and non-operating funds. Operating excesses result when operating revenues exceed operating costs. Utilities must have a certain amount of excess cash for working capital purposes. In a regulated environment, most public service regulatory agencies recommend a working capital reserve of approximately one-eighth to one-sixth of total revenues. A higher working capital fund can be justified by conducting a "lead-lag" study. Most rate technicians would argue that any excess above a proper working capital reserve could be used as an offset to revenue requirements.

The use of interest on special reserve funds for capital-related requirements is less clear. Most of these funds are established by bond ordinance with the interest earnings allowed to be used for specified purposes such as (1) offsetting revenue requirements, (2) offsetting coverage requirements, or (3) any utility purpose the utility deems appropriate.

Special capital fees and charges are related to specific capital pro-grams or facilities and would not normally be used for offsetting revenue requirements to be recovered from user charges. A good example of such a fee would be an impact fee, as discussed in Chapter 4. In most instances, impact fees are restricted for funding capital improvements related to expansion projects. In some cases, however, bond ordinances will allow impact fees to be used in debt service coverage calculations; therefore, they can be used to offset the debt service portion of user charge revenue requirements.

Sales of assets are usually unplanned. Excess machinery, vehicles, and inventory can be sold and the revenues used to offset large reve-nue requirements.

An effective chart of accounts should track specific service charge revenues and revenues from non-operating sources. The revenue accounts would then provide a basis for calculating the offset to revenue requirements.

B. Allocation of Water Cost of Service

After removing revenues from specific service charges and non-operating sources from revenue requirements, it is necessary to allo-cate net revenue requirements to classes of customers. This allocation process usually takes place in two steps: (1) allocation of costs to functional cost-of-service categories, and (2) reallocation of func-tional costs to classes of customers.

As depicted in Figure 9.3, water costs (net revenue requirements) would normally first be allocated to functional cost-of-service components:

- *Source of supply:* operating and capital costs associated with the source of water supply (reservoir construction and maintenance costs, water right purchases, supply development costs, conservation costs, etc.)

- *Pumping and conveyance:* costs associated with pumping raw water from the source of supply and transferring it through a piping network for treatment

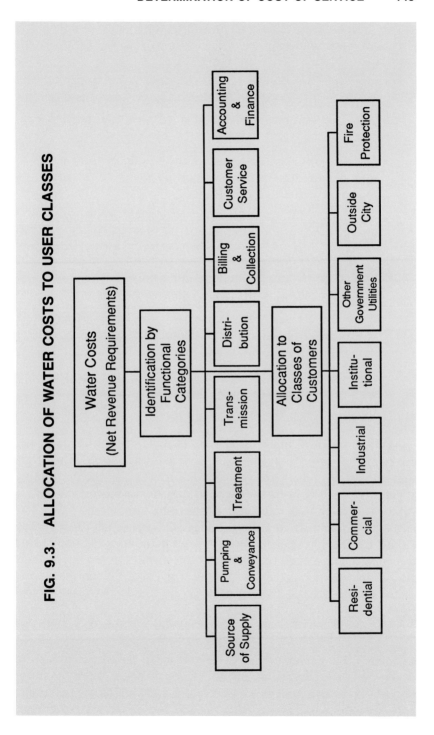

FIG. 9.3. ALLOCATION OF WATER COSTS TO USER CLASSES

- *Treatment:* costs associated with treating water to potable water standards

- *Transmission:* costs associated with transporting water from the point of treatment through a major trunk to major locations within the service area. Water storage costs are normally considered a part of transmission costs.

- *Distribution:* costs associated with the smaller local service distribution mains transporting water to specific locations within the service area

- *Administration:* overhead costs associated with managing the water utility operations

- *Billing and collection:* meter reading, billing, and collection costs associated with preparing a water customer bill and processing funds received from water users

- *Customer service:* costs associated with administering customer accounts (processing complaints, responding to customer inquiries, performing rereads, etc.)

- *Accounting and finance:* administrative costs associated with accounting, reporting, investment processing, and other financial activities

As discussed in Chapter 8, if the NARUC chart of accounts is effectively integrated into the utility's accounting system, identification of cost by functional category is provided by the accounting system. If the accounting system does not provide this breakdown, it will be necessary to develop allocations using appropriate bases. For example, "Planning and Capital Budgeting" costs could be allocated among functional categories based upon the ratios of net book value of capital items related to functional categories. An example of this allocation process is presented in Table 9.1. This table translates costs under a governmental accounting structure to a NARUC cost structure.

Some pricing methodologies would require that functional costs be reallocated to other cost categories before being allocated to classes of customers. For example, under the base–extra capacity approach to water rate setting identified by the American Water Works Association (AWWA) as a preferred cost allocation methodology, functional costs are allocated to average day, maximum day, and maximum hour components before being reallocated to customer classes. Under another AWWA-accepted methodology, the demand-commodity approach, costs are allocated to demand and commodity costs before being reallocated to classes of customers.

After water cost pools are established for cost-of-service purposes, it is necessary to allocate each cost pool of functional cost to classes of water customers. A customer class could be established if their specific loading characteristics are somewhat different than other classes of users. In other words, a customer class should be established if the cost to serve that particular class of customer is different than other classes of customers.

Several factors that differentiate the cost of providing service among customer classes include:

- *Demand characteristics*: the ratio of peak usage to average usage by a class of customer. For example, the ratio of maximum hour usage to average day usage could vary significantly from a residential class to an industrial class. As a result, customers with high peaking factors require more cost to service on a per-unit basis than customers with low peaking factors. In other words, the load of one customer class may be more efficient than that of another class. As discussed above, the base–extra capacity approach to rate setting recognizes demand and load characteristics as the major basis for differentiating cost of service among classes of customers.

- *Location of customers:* The distance that a customer is from the potable water supply makes a difference in the cost of delivering water to a particular class of customer. Some utilities have elected to establish "districts" or "pressure zones" for cost-of-service and rate-setting purposes.

- *Types of mains serving specific customer classes:* In some cases, only the larger transmission mains are used to serve large-volume customers such as industries, institutional customers, and other governmental utilities. It can be argued that these customers should only share in the operating and capital costs of the larger transmission system and not recover costs related to the smaller distribution lines serving primarily residential and other small-volume customers.

Table 9.1　Allocation to NARUC Functional Categories

Expense Category	Total Amount (In $1000's)	Source of Supply— Operations	Source of Supply— Maintenance	Water Treatment— Operations	Water Treatment— Maintenance
Administrative Departments					
• Billing & Collection & Customer Service	$ 1,000				
• Data Processing	500				
• Accounting	300				
• Insurance & Risk Management	100				
• Personnel	200	20	15	50	45
• Purchasing	150	5	5	15	30
• Grants Administration	200	100	80	20	
TOTAL ADMINISTRATION	2,450	125	100	85	75
Engineering Department					
• Planning & Capital Budgeting	500	50	37.5	125	112.5
• Construction Inspection	500				
• Demand Forecasting	200	20	15	50	45
• Developer Review	350				
• Project Design	800				100
TOTAL ENGINEERING	2,350	70	52.5	175	257.5
Operations Department					
• Administration	300	15	15	80	50
• Construction	2,500			200	300
• Leak Detection	300				
• Treatment	5,500			4,500	1,000
• Service Installation	1,000				
• Line Maintenance	2,500				
• Laboratory	800			800	
• Pumping	1,200	1,200			
TOTAL OPERATIONS	14,100	1,215	15	5,580	1,350
TOTAL O&M COSTS	$18,900	1,410	167.5	5,840	1,682.5

• *Age of facility:* Many times, existing customers argue that they should not have to pay for costly expansion facilities that benefit new customers and growth. Even though it may be rare, some rate structures attempt to match certain facilities with a certain class of customers and, therefore, justify a cost differential.

Examples of classes of customers that have been established in water rate systems include:

- residential
- commercial (break down by category, if appropriate)
- industrial (break down by category, if appropriate)
- institutional (colleges, schools, hospitals, etc.)

Table 9.1   Continued

Trans. & Dist. Operations	Trans. & Dist. Maintenance	Customer Account	Admin. and General	Basis for Allocation
		$1,000		Direct Identification
		300	200	Direct Identification
		50	250	Direct Identification
			100	Capital Net Book Value by Functional Category
15	20	15	20	Headcount by Operational and Administrative Unit
20	25	20	30	Number of Purchase Orders
				Grant Amounts by Functional Area
35	45	1,385	600	
37.5	50	37.5	50	Capital Net Book Value by Functional Category
300	200			Work Orders
15	20	15	20	Capital Net Book Value by Functional Category
200	150			Work Orders
400	300			Time Sheets
952.5	720	52.5	70	
50	50	20	20	Direct Identification
1,500	500			Work Orders
	300			Direct Identification
				Direct Identification
		1,000		Direct Identification
	2,500			Direct Identification
				Direct Identification
				Direct Identification
1,550	3,350	1,020	20	
2,537.5	4,115	2,457.5	690	

- other governmental utilities (each utility is usually considered separately)
- outside city customers
- fire protection
- other classes as deemed appropriate by local cost-of-service circumstances

Fire protection deserves special mention because the utility may choose to recover fire protection costs through water rates, an interfund transfer from the city/county general fund, charges to special fire districts, and/or private fire line fees.

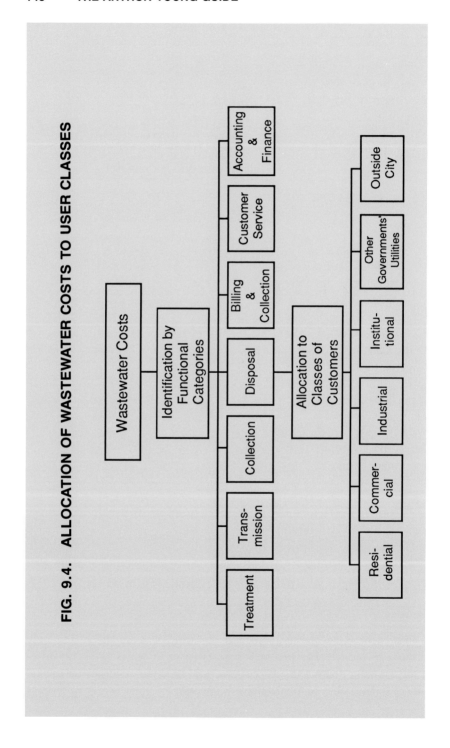

FIG. 9.4.   ALLOCATION OF WASTEWATER COSTS TO USER CLASSES

C. Allocation of Wastewater Cost of Service

The allocation of wastewater costs (net revenue requirements) to cost-of-service categories is performed in a manner similar to the allocation of water costs. As depicted in Figure 9.4, functional cost categories are modified to reflect the operational nature of the wastewater utility.

Functional categories could include:

- treatment
- transmission
- collection
- disposal
- administration
- billing and collection
- customer service
- accounting and finance

Customer classes would likely be similar to water customer classes. In the case of wastewater cost of service, special consideration would be given to differences in wastewater strength and infiltration/inflow (I/I) among classes of customers. For wastewater strength, special cost allocations could be made for the treatment of special pollutants such as biological oxygen demand (BOD), chemical oxygen demand (COD), suspended solids (SS), ammonia, and phosphorus. The degree of I/I associated with specific lines serving certain classes of customers would be considered.

10   DESIGNING A WATER AND WASTEWATER RATE STRUCTURE

10 DESIGNING A WATER AND WASTEWATER RATE STRUCTURE

After costs have been allocated to customer classes, the next step is to design a rate structure which will recover costs appropriately from these classes of customers. Several key issues have to be addressed in designing an appropriate rate structure:

- Should a special rate structure be established for each customer class?

- Should a combined rate structure be developed which could be used by all customer classes?

- Should there be a minimum (or fixed) charge component and a consumption (or variable) charge component to the rate structure?

- If a minimum charge is desired, what elements of cost should be included in the minimum charge?

- If a combined rate structure is desirable, how many usage blocks should there be and what should the cut-off point for each block be?

Figure 10.1 depicts the rate structure process for water rates. The goal of an effective rate structure is to relate the cost of service to customer classes while maintaining simplicity and ease of implementation. In addition, the rate design should identify those costs that should be recovered through fixed charges and consumption charges.

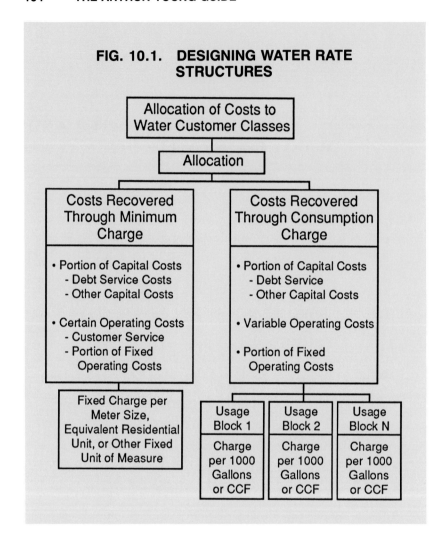

FIG. 10.1. DESIGNING WATER RATE STRUCTURES

A. Establishing an Appropriate Minimum (or Service) Charge

The rationale for having a minimum charge is to recover certain costs as a fixed component of the customer's bill. The more costs recovered through the minimum charge, the more guaranteed revenue the utility can expect. On the other hand, the less control that the customer has in affecting his ultimate charge (as with a high minimum charge), the less likely he will be to conserve usage. Bond rating agencies look favorably upon user charge structures that recover a high percentage of revenue requirements through fixed charges, since bond holders are more protected when revenue is less dependent upon usage.

Some rate technicians assume that a minimum charge provides for some allowance of water or wastewater usage. The term "minimum charge" in the context of this book does not require the charge to include such an allowance. The minimum charge as discussed in this chapter is defined as a fixed "service charge" that might or might not include a usage allowance. Several types of costs could logically be recovered through the minimum charge. First of all, capital costs associated with facilities that are available for providing basic service to the customer could appropriately be recovered through the minimum charge. When debt is used to finance major facilities, the utility has to pay debt service whether usage materializes or not. By recovering debt service costs through the minimum charge, the utility would be passing this fixed cost proportionately to each customer.

Another type of cost that could logically be recovered through the minimum charge would be customer service costs. These costs include the costs associated with (1) servicing a customer's account (billing, collection, meter reading, and customer service costs, approximately equal for each customer account) and (2) those related to meter installation, testing, and maintenance (typically varying by meter size).

Finally, an argument can also be made for including other fixed operating costs in the minimum charge. Certain fixed operating costs have to be paid by the utility whether or not usage materializes, and the utility could logically recover these costs through the minimum charge. In an extreme example, all utility costs might be recovered through the minumum charge. Such structures are called "flat rate" systems. In determining the types of cost to be recovered through the minimum charge, the utility would need to evaluate the impact on customers and other pricing objectives.

After the costs to be recovered through the minimum charge are identified, the next step is to identify the appropriate unit of measure for recovering these costs. The unit of measure may vary for different types of costs. For example, the utility may desire to recover customer service costs on a "per account" basis. This recovery approach assumes that it takes approximately the same amount to service each customer's account. On the other hand, availability costs, or capital-related costs, might better be allocated on meter-size equivalents. The meter size represents a potential level of demand placed on the water system by the customer, with the larger meter size recovering a greater percentage of demand-related cost. Another typical unit of measure would be the equivalent residential unit measure. Under the ERU approach, fixed charges would be based upon the ratio of a customer's reading expressed in terms of a typical residential customer. These measures are the same as would be considered in assessing capital recovery charges, and are discussed in detail in Chapter 4.

B. Establishing an Appropriate Consumption Charge

Costs not recovered through minimum charges must be recovered through consumption (water) or usage (wastewater) charges. Charges based upon consumption vary by the amount of usage, typically measured through water meter readings. In the case of wastewater, some percentage of water consumption might be used in estimating wastewater discharged by the customer. A simplistic approach to developing a consumption charge would be to simply divide total cost by total projected billable consumption, resulting in a charge per thousand gallons or hundred cubic feet (CCF). In this case, there would be only one usage block and all customers would be billed based upon this uniform rate.

Rate structures become much more complex when the usage blocks are structured to reflect cost-of-service characteristics of different classes of customers. In a multiblock rate structure, it is the goal of the rate technician to establish usage blocks and related rates which reflect the cost of service of different customer classes. In a well-designed rate structure, each block will have primarily the cost-of-service characteristics of one class of customer. There is an exception to this philosophy if the rate blocks are based on non-cost principles (some conservation-based structures).

As a first step in developing appropriate blocks, a bill frequency analysis is prepared for each class of customer. A bill frequency analysis will identify how much consumption a particular customer class uses in certain billing intervals and what percentage of customer bills fall within the billing interval. An example of a bill frequency analysis for the residential class of customers is presented in Table 10.1.

Based upon the results of the bill frequency analysis, a usage curve for each class of user is plotted, typically on a logarithmic graph. The rate technician then determines usage levels at which (1) the majority of usage of one particular class of customer is captured, and at the same time, (2) a minimum amount of usage from other customer classes is recovered. In Figure 10.2, the bill frequency data for the

Table 10.1   Block Analysis of Monthly Consumption for Residential Class

Consumption Amount (in CCF)	# of Bills Terminating in Interval	% of Total Bills	Cumulative % of Total Bills	Consumption in Interval (in CCF)	% of Total consumption	Cumulative % of Total Consumption
0	400	1%	1%	0	0	0
1	500	1	1	300		0
2	1,500	4	6	2,250	1	1
3	2,100	5	11	5,800	1	2
4	2,400	6	17	8,800	2	4
5	2,500	6	23	11,500	3	7
6	2,900	7	30	16,200	4	11
7	3,500	9	39	22,750	5	16
8	4,000	10	49	31,600	7	23
9	3,500	9	58	31,200	7	30
10	3,500	9	67	32,900	7	37
12	2,600	6	73	29,100	7	44
14	2,200	5	78	30,400	7	51
16	1,900	5	83	29,600	7	58
18	1,500	4	87	26,700	6	64
20	1,300	3	90	24,500	6	70
24	1,100	3	93	24,200	6	76
28	900	2	95	24,300	6	82
32	700	2	97	21,700	5	87
36	500	1	98	17,000	4	91
40	300	1	99	11,800	3	94
<40	300	1	100	27,400	6	100
TOTAL	40,100	100%		430,000	100	

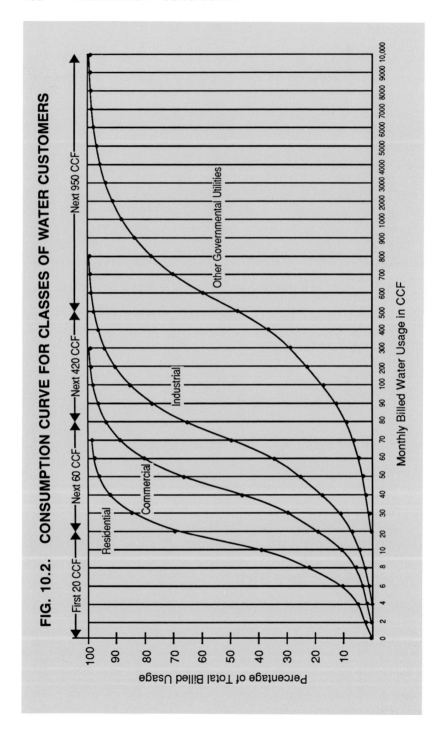

FIG. 10.2.   CONSUMPTION CURVE FOR CLASSES OF WATER CUSTOMERS

residential customer class (presented in Table 10.1) is plotted on a logarithmic graph. In addition, the bill frequency consumption curve is plotted for the customer classes of commercial, industrial, and other government utilities. The majority of residential consumption occurs in usage block 0 to 20 CCF. At the same time, a minimum of commercial, industrial, and other governmental utility consumption occurs. As a result, the charge for rate block 0 to 20 CCF should reflect predominately the residential class unit cost of service. As depicted in Figure 10.2, the consumption pattern of each class of customer should be plotted on the same graph in order to facilitate identifying the appropriate cut-off points for each usage block. Once this process is completed, there should be approximately the same number of usage blocks as there are classes of customers.

The unit rate for each block is calculated by taking the individual cost-of-service rates for each class of customer having consumption within a block, weighted by the consumption of each class of customer within the block. For example, the rate for Block 2 (from 20 CCF to 80 CCF), as presented in Table 10.2, would be $0.94 per CCF. The Block 2 rate would have the predominant characteristics of the cost-of-service rate for the commercial class of customer ($0.95/CCF).

Note that if cost-of-service rates were charged each class of customer, Block 2 revenue would be $30,680 ($1.10 x 2500 CCF + $0.95 x 27,000 CCF + $0.70 x 3000 CCF + $0.60 x 300 CCF); this is the same approximate outcome ($30,832) as charging a "blended" cost of service rate of $0.94 for all consumption in Block 2.

Rate blocks will not normally fall out as nicely as they did this example. Consumption patterns among customer classes can be complicated. Experience and judgment is important in determining where the cut-offs for each block should be established.

After all usage blocks are established, a revenue sufficiency test should be performed by multiplying rates and charges by expected usage levels to confirm that adequate revenues will be generated, assuming usage occurs as planned.

C. Water Rate Structure Examples

There are several rate structures that are widely used across the country by government utilities. Each rate structure is justified by a set of engineering, accounting, and/or economic principles with each

Table 10.2   Calculation of Rate for Block 20 to 80 CCF

Customer Class	# of Customers	Consumption in Block 20 to 80 CCF	% of Total	Cost of Service Unit Cost	Weighted Block Charge[a]
Residential	10,000	2,500	8	$1.10	$.088
Commercial	500	27,000	82	.95	.779
Industrial	50	3,000	9	.70	.063
Other Gov't Utities	5	300	1	.60	.006
TOTAL	10,555	32,800	100%		.936
					$.94/CCF (Rounded)

[a]Column 4 times Column 5.

Table 10.3   Rate Structure Examples

Rate Structure Type	Monthly Minimum Charge		Monthly Consumption Charge
	Meter Size	Charge	
Demand-Commodity Approach	5/8″	$ 3.15	
	3/4″	4.75	
	1¹/₂″	12.35	$1.35 per thousand gallons
	2″	22.50	
	3″	42.85	
	(etc.)		
Customer-Commodity Approach	5/8″	$ 2.00	
	3/4″	2.00	
	1¹/₂″	2.00	$1.45 per thousand gallons
	2″	2.00	
	3″	2.00	
	(etc.)		
Base–Extra Capacity Approach	5/8″	$ 3.15	(1000 gallons)
	3/4″	4.75	0–10 $1.65 per thousand
	1¹/₂″	12.35	11–100 1.35 gallons
	2″	22.50	101–1000 1.10
	3″	42.85	> 1000 .85
	(etc.)		
Marginal Pricing	Similar to Base–Extra Capacity Approach		$2.00 per thousand gallons
Conservation Pricing	Similar to Base–Extra Capacity Approach		(1000 gallons)
			0–10 $1.35 per thousand
			11–100 1.60 gallons
			101–1000 1.85
			> 1000 2.00

structure having a specific impact on different water and wastewater customers. Rate structure examples are presented in Table 10.3.

D. Establishing Wastewater Rates

In establishing a rate structure for wastewater, the process is similar. The major difference is that wastewater strength plays an important part in establishing charges for users discharging wastewater with strengths greater than sanitary sewage. In such cases, a surcharge would be assessed to customers on the basis of how much they exceed the sanitary sewage level for certain pollutants. The surcharge is normally expressed in terms of pounds of pollutant that is being surcharged.

EPA user charge regulations also affect how wastewater rates are structured. Any government utility that has received wastewater grant funds under the Clean Water Act must develop an approved user charge system. These regulations generally require that wastewater operating, maintenance, and replacement (OM&R) costs be recovered proportionately from each user or class of user based upon cost of service. EPA user charge regulations do not address non-OM&R costs such as debt service or pay-as-you-go capital costs. *If the rate-setting principles in this chapter are followed, a government utility will comply with EPA user charge regulations.*

E. Selecting the Appropriate Rate Structure

It is important to evaluate the pricing philosophy of the community in selecting the appropriate rate structure. Factors that could be considered include:

- financial sufficiency
- equity
- rate stability
- impact on customers
- understandability
- simplicity
- conservation
- legality

Typically, a process similar to the process used in selecting an appropriate financing plan (discussed in Chapter 6) should be followed.

11 COMPARING WATER AND WASTEWATER RATES AMONG UTILITIES

11   COMPARING WATER AND WASTEWATER RATES AMONG UTILITIES

For 1988, Arthur Young's National Environmental Consulting Group conducted a survey comparing water and wastewater rates among the 120 largest communities throughout the United States. The survey targeted the major water and wastewater utilities serving the central and, as applicable, surrounding cities of the 100 largest Metropolitan Statistical Areas (MSAs). The MSAs were ranked according to population estimates developed by the U.S. Office of Management and Budget as of June 30, 1985. During the survey, 120 communities and 161 utilities were contacted. Of the utilities contacted, 153 utilities, or 95%, responded. In some communities, separate water and wastewater utilities served customers.

A. Survey Results

The survey results were compiled during the fall of 1987. Appendix C (monthly water charges) and Appendix D (monthly wastewater charges) summarize survey results by presenting the inside-city charges for each community at various consumption levels based upon the most common meter used at that level. Rates were calculated at monthly usage levels; the survey averages for each consumption level are presented in Table 11.1.

The meter size presented is a typical meter size for a customer with the relevant amount of usage.

To demonstrate the variations in charges, Figures 11.1 and 11.2 compare the lowest, average, and highest water and wastewater charges among surveyed utilities. The charges are calculated for a typical customer using 10 CCF or 7,480 gallons of service.

Appendix E describes the water and wastewater rate structure, identifies when each utility last updated its rates, and includes the

Table 11.1   Survey Averages for Each Consumption Level

Monthly Usage in Cubic Feet	Monthly Usage in Gallons	Meter Size Inches	Monthly Water Bill	Monthly Waste-Water Bill
0	0	$5/8$	$ 3.32	$ 3.78
500	3,750	$5/8$	6.12	6.71
1,000	7,480	$5/8$	9.95	10.84
3,000	22,440	$5/8$	26.09	26.84
50,000	374,000	2	355.47	441.80
1,000,000	7,480,000	4	6,275.29	8,502.34
1,500,000	11,200,000	8	9,507.20	12,784.12

frequency of billing for each respondent. A summary of the types of water rate structures by region of the country is presented in Table 11.2. The percentage of water rate structures by type is depicted in Figure 11.3.

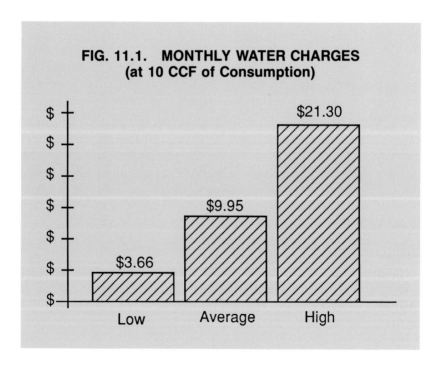

FIG. 11.1.   MONTHLY WATER CHARGES
(at 10 CCF of Consumption)

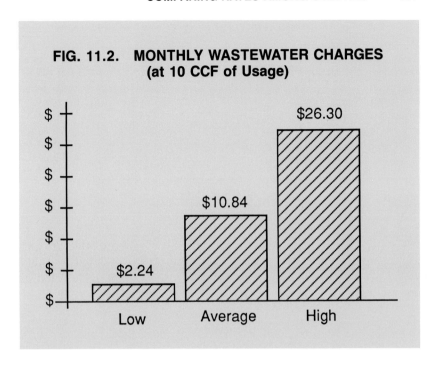

FIG. 11.2. MONTHLY WASTEWATER CHARGES (at 10 CCF of Usage)

B. Comparing Rates Among Communities

Comparing water and wastewater rates among communities can provide insights into pricing policies of water and wastewater utilities. Care should be taken, however, in drawing conclusions from this comparison. High rates may not mean the utilities are operated and managed poorly. Many factors affect (1) costs to be recovered through water and wastewater pricing and (2) the pricing structure.

Table 11.2   Water Rate Structures Used by Rate Survey Respondents

Rate Structure	Northeast	Southeast	Southwest	Midwest	West	Total
Declining Block	12	16	5	22	2	57
Uniform	5	8	7	4	12	36
Inverted Block	2	4	5	2	6	19
Total	19	28	17	28	20	112

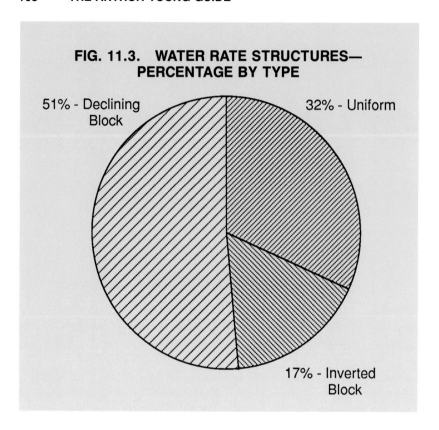

FIG. 11.3.   WATER RATE STRUCTURES—
PERCENTAGE BY TYPE

Some of the most prevalent factors include:

• geographic location
• demand
• customer constituency
• level of treatment
• level of general fund subsidization
• level of grant funding
• age of system
• infiltration and inflow problems
• rate setting methodology
• other evaluation criteria

These factors are discussed below.

1. GEOGRAPHIC LOCATION

Geographic location (urban/rural) and topography significantly affect the design and cost of water and wastewater facilities and their operation. In some areas, pumping and transmission costs can be major system costs. Water treatment plants located far from the source of water supply can have high water supply costs. A wastewater treatment plant located far from its discharge stream can have high disposal costs.

Another geographical consideration is customer density. In areas where water and wastewater connections are dense, collection and distribution costs can be significantly lower than in areas where customers are more spread out.

2. DEMAND

Customer demand plays an important role in sizing water and wastewater facilities, and therefore affects water and wastewater rates. Facilities have to be designed to provide for seasonal and hourly demand, as well as potential growth in a system. Peak demand usage may be significantly higher than average annual usage of water and wastewater facilities. As a result, customers may have to pay a relatively higher rate during nonpeak periods to have facilities available to be used during peak periods.

Resort areas provide a good example of the impact of peak demand on water and wastewater pricing. Facilities are sized to meet vacation demand and have high facility costs where computed on an average annual gallon basis. Communities which maintain stringent fire protection standards might have relatively high peak hour water demands, and therefore incur additional operating and facility costs related to providing fire protection. Many jurisdictions, however, recover fire protection costs through charges to either the city's or the county's general fund or to special fire districts with taxing authority. In these cases, the water customer rate base can be relieved of recovering the cost to provide fire protection.

3. CUSTOMER CONSTITUENCY

The types of customers served by a water or wastewater system affect administrative, customer service, treatment, and transmission costs. In communities with numerous high-volume users, administrative, customer service, and transmission costs can be relatively low.

Factors contributing to this lower rate include (1) more gallons can be consumed or discharged per foot of line, (2) fewer meters need to be read and bills prepared, and (3) less administration is involved with delinquencies, disconnects, and customer service. On the other hand, areas with high industrial discharge can incur significantly more operating and capital costs to (1) treat and process wastewater, (2) maintain an industrial waste control or pretreatment section, and (3) provide for more expensive metering equipment.

4. LEVEL OF TREATMENT

A wastewater plant's effluent quality standards are established by the state and identified in the plant's National Pollution Discharge Elimination System permit. These standards are influenced by the water quality of the discharge stream, as well as the pollutants that must be treated. The level and type of wastewater treatment influences wastewater treatment design and related operating and capital costs. Communities with tertiary treatment would typically incur greater costs than communities served by secondary treatment plants.

For water treatment, the quality of influent water affects treatment costs. In many situations, groundwater is relatively pollutant-free and can be distributed after little treatment. Treatment of surface water would typically be more complicated, and therefore more costly.

5. LEVEL OF GENERAL FUND SUBSIDIZATION

Many public water and wastewater operations are organizationally within a county or municipal government. The county or municipal government often provides administrative services which benefit water or wastewater operations. These services might include personnel services, purchasing, administration, accounting, and data processing. If the general fund does not recover sufficient administrative costs from water or sewer operations, a subsidy to water or sewer operations could result. On the other hand, over-recovery of administrative costs from water and wastewater operations could result in a subsidy to the general fund. The subsidy concept could be extended to include subsidizing more than administrative costs.

In the case of water and wastewater authorities, the authorities' operations are self-contained and no "parent" governmental body exists to create a subsidy situation.

6. LEVEL OF GRANT FUNDING

Grant funding from state and federal agencies can be an offset to water and wastewater capital costs and ultimately water and wastewater rates. However, "201" wastewater construction grants are declining as the percentages of participation are reduced. Federal water grant assistance has almost disappeared.

In comparing rates, one would think that grant funding would have a neutral impact on all communities receiving grant funds. This is not necessarily true, however, since (1) each area may have a different level of projects eligible for funding, and (2) some states supplement federal funding with a state match. As a result, the local share can be significantly different from community to community, and rates will be affected accordingly.

In the case of grant funding for water projects, some communities have received state water grants or special federal assistance (Farmers Home Administration, Economic Development Administration, etc.). Again, the level of water funding would impact water capital requirements, and the level of capital revenue requirements to be recovered from water customers.

7. AGE OF THE SYSTEM

Typically, older systems require more maintenance. In new treatment systems, however, an extraordinary amount of maintenance may be required to work out the "bugs" until the facilities are operating efficiently. Also, with a new system, significant debt service costs may be required, as compared with older systems where debt has been repaid, or where the debt is based on much lower service costs. As a result, the age of the system should be evaluated to determine operating and capital revenue requirements and the impact on rate comparisons.

8. INFILTRATION AND INFLOW LEVELS

A major problem with many wastewater systems is the level of infiltration and inflow (I/I) present. A high level of I/I means additional capacity requirements and related operating costs. These additional costs translate into higher revenue requirements in the rate base.

9. RATE-SETTING METHODOLOGY

The methodology used by a community to establish rates influences (1) the level of revenue requirements to be recovered through rates, and (2) the distribution of costs to classes of water and wastewater customers. If the "cash-needs" approach to rate setting is followed, cash requirements for utility operations are recovered through the rate base. If the "utility" approach is followed, accrual revenue requirements are recovered through the rates. The major difference between revenue requirements under the two approaches is the handling of capital costs. Under the cash-needs approach, debt service with related cash requirements and other capital cash outlays are recovered through the rates. Under the utility approach, depreciation and rate of return are recovered through the rates. In a climate of significant grant funding, capital recovery requirements could be significantly different under the two approaches.

The method used to allocate costs to customer classes and related consumption blocks influences the charges at different consumption levels. If a uniform rate approach is followed, each unit of service (thousand gallons, CCF, etc.) is charged at the same price for all levels of usage. Under a declining block rate structure, the charge per unit of service is lower at the higher usage levels than at the lower usage levels. Under this methodology, charges for the first CCF of usage would be greater than at high volume levels. The inverted block rate structure is designed to promote conservation and presents another cost allocation method. Under this approach, unit charges increase as consumption increases. The rate-setting approach adopted by a community has a significant impact on the way costs are distributed among the consumption levels and classes of users.

10. OTHER EVALUATION CRITERIA

Other factors influencing the comparison of rates are too numerous to mention. These factors would relate to levels of efficiency, organizational considerations, and demographic considerations, such as availability of labor, compensation scales, and levels of employee training.

In summary, care should be taken in drawing conclusions regarding water and wastewater operations or maintenance in a particular community. Many factors influence water and wastewater pricing. However, comparisons among communities could signal to management that there are reasons why one community's rates are higher or lower than another community's. Analysis of why there is a difference could be helpful in examining the effectiveness of a water or wastewater operation.

APPENDIXES

A

CALCULATING CAPITAL RECOVERY CHARGES UNDER ALTERNATIVE APPROACHES

CALCULATING CAPITAL RECOVERY CHARGES UNDER ALTERNATIVE APPROACHES

BACKGROUND

The City of Georgetown is a medium-sized community of 55,000 residents. The city is in the process of expanding and upgrading its water treatment and distribution system to address economic development and the Safe Drinking Water Act, and wants to ensure that new customers pay for expansion of related capital costs and that all customers share in the cost of upgrading. The city utility director would like to implement a fair and equitable capital recovery charge structure. He has asked his finance director and city engineer to assist him in developing the charge under alternative approaches. Relevant information that is available to the finance director and the city engineer is shown in the following sections.

GENERAL INFORMATION

Number of water customers — 21,500
Equivalent residential units (ERU) — 28,900
Usage per ERU — 280 gallons per day

Five-Year Water Capital Improvement Plan (CIP) (in $1,000's)

Capital Component	FY1989	FY1990	FY1991	FY1992	FY1993	FY1994
Source of Supply	$1,000		$ 500		$ 700	$2,200
Treatment and						
Pumping	4,000	$1,100		$ 400		5,500[a]
Transmission	1,500	1,600				3,100
Distribution	250	300	350	300	350	1,550
Services, Meters,						
and Hydrants	600	500	950	200	600	2,850
General Structures		1,000				1,000
	$7,350	$4,500	$1,800	$ 900	$1,650	$16,200

[a]Nature of treatment and pumping facilities: upgrade—$1,100,000; expansion—$4,400,000.
Additional capacity from CIP: 2,900,000 gallons per day.
Project financing of CIP: $17,500,000 revenue bond to be issued in FY1989, 20-year amortization period at 10% interest.
All amounts expressed in FY1989 dollars.

WATER CAPITAL STRUCTURE

Capital Component	Net Book Value (in $1,000's)
Source of Supply	$2,100
Treatment and Pumping	9,200
Transmission	5,100
Distribution	5,400
Services, Meters, and Hydrants	4,800
General Structures	2,700

Outstanding debt: $10,500,000 (revenue bonds).
Developer contributions: $2,100,000 (water treatment facility).

The finance director and the city engineer calculated capital recovery charges under the following approaches:

- growth-related cost allocation method
- marginal-incremental cost method
- system buy-in approach
- value-of-service method

1. Growth-Related Cost Allocation Method

Projected capital improvement program	$16,200,000
Less: capital costs related to	
treatment plant upgrade	(1,100,000)
distribution facilities	(1,550,000)
service, meters, and hydrants	(2,850,000)

Net expansion costs of backup system	$10,700,000
Expansion capacity	2.9 MGD
Capital recovery unit costs	$3.69 per gallon
Capital recovery charge per ERU at 280 gallons per day	$1,033

2. Marginal-Incremental Cost Method

Revenue bond amount for CIP	$17,500,000
Less: Portion of debt related to treatment upgrade and distribution, and on services, meters, and hydrants ([$5,500,000 ÷ $16,200,000 = 34%] × $17,500,000)	(5,950,000)
Portion of bond user charge revenue stream will support	(800,000)
Bond amount to be recovered by capital recovery charge	$10,750,000
Additional capacity from CIP program	9,670 ERUs
Capital recovery charge per ERU	$ 1,112

3. System Buy-In Approach

Capital Component	Net Book Value Adjusted to 1989
Source of Supply	$ 6,100,000
Treatment and Pumping	19,400,000
Transmission	8,900,000
Distribution	10,100,000
Services, Meters, and Hydrants	7,400,000
General Structures	7,300,000
	$59,200,000
Less: Outstanding Debt	$(10,500,000)
Developer Contributions (Net Book Value Adjusted to FY1989 Dollars)	(4,300,000)
Distribution Facilities	(10,100,000)
Service, Meters, and Hydrants	(7,400,000)
Total	$26,900,000
Existing Capacity	28,900 ERUs
Capital Recovery Charge	$931

4. Value-of-Service Method

The finance director and city engineer surveyed several government utilities with similar operational and capital requirements. They determined that the average water capital recovery charge was $850. They also priced a well system for a single-family dwelling at $1,500. Using the value-of-service method, they determined a capital recovery charge of $1,000. We emphasize that using the value-of-service method alone in establishing capital recovery charges is dangerous and has a high litigation potential.

Capital recovery charges developed under alternative approaches are summarized below:

Method	*Charge per ERU*
Growth-related cost allocation method	$1,033
Marginal-incremental cost method	1,112
System buy-in method	931
Value-of-service method	1,000

As a result of their analysis, the finance director and city engineer felt that the growth-related cost allocation method most appropriately reflected their capital recovery charge philosophy. As a result, they recommended a capital recovery charge of $1,033 per ERU.

B

TOTAL ONE-TIME CHARGES ASSESSED FOR A NEW SINGLE-FAMILY RESIDENCE TO CONNECT TO THE WATER/WASTEWATER SYSTEM

Source: Arthur Young's 1988 National Water and Wastewater Rate Survey.

1988 National Water/Wastewater Rate Survey: Total One-Time Charges

	Total One-Time Charges Assessed for a New Single-Family Residence to Connect to the Water/Wastewater System			
State/Cities	Water	Description	Wastewater	Description
ALABAMA				
Birmingham	$135.00	Tap fee	Varies	
Mobile	225.00	Minimum for 60' front, all additional feet at $3.75 per foot.	$450.00	Minimum for 60' front, all additional feet at $7.50 per foot.
ARIZONA				
Phoenix	Varies		Varies	
Tucson	370.00	Plus Area Development Fee (ADF), if applicable. ADF range: $188–$405.	599.78	Sewer connection fees based on 18 fixture units for participating property plus sewer fee turn-on charge of $10
ARKANSAS				
Little Rock	120.00		100.00	Plus $10 inspection fee.
CALIFORNIA				
Anaheim	Varies	$5.11 per 100 sq. ft. of land area plus $1,120 for 5/8" meter and 1" service tap. Other fees dependent on location.	Not available	
Bakersfield	None		662.50	
Fresno	1,125.00		678.00	
Los Angeles	1,720.00		1,261.50	
Oakland	950.00–5,480.00	Depending on region and type of installation.	390.00	
Sacramento	2,146.00	Development fee: $1,500 Tapping fee: $646	1,100.00	Average tapping fee.
San Diego	565.00		1,483.00	
San Francisco	1,200.00		0.00	Note: Service must be installed by private contractor.

			Minimum	
San Jose	0.00	(San Jose Water Company)	Minimum: 1,465.00	Charge comprised of sanitary connection fee, storm drainage fee, and sewage treatment plant connection fee.
Stockton	150.00		1,600.00	
Ventura	1,318.00		679.00	
COLORADO				
Colorado Springs	3,807.00	System development charge (3/4").	619.00	
Denver	2,730.00		287.00	
CONNECTICUT				
Hartford	Varies	Per front foot: $21.25 High service per foot: 1.00 Tapping fee: 76.50	Varies	Per lateral: $640.00 Per front foot: 27.85 Rate per acre: 860.00
New Haven	485.00		0.00	
DISTRICT OF COLUMBIA				
Washington	78.00		0.00	
FLORIDA				
Fort Lauderdale	408.00		0.00	
Jacksonville	270.00	5/8" and 3/4" meter.	1,070.00	
Lakeland	Varies		90.00	Tap fee only; additional charges for water pollution, high strength impacts.
Miami	315.00	With existing box.	301.00	
	815.00	Without existing box.		
Orlando	435.00		2,550.00	
St. Petersburg	160.00		350.00	
Tampa	1,300.00		800.00	
West Palm Beach	400.00		1,700.00	Per water closet.

Total One-Time Charges Assessed for a New Single-Family Residence to Connect to the Water/Wastewater System

State/Cities	Water	Description	Wastewater	Description
GEORGIA				
Atlanta	$440.00	Short side, plus varying cost for sidewalk and streetcutting.	$0.00	
	620.00	Long side, plus varying cost for sidewalk and streetcutting (³/₄″ meter charges, more for larger sizes.)		
Augusta	Not available		Not available	
HAWAII				
Honolulu	1,545.00		0.16	Per square foot (residential).
ILLINOIS				
Chicago	315.00		45.00	
Joliet	110.00		110.00	
Peoria	0.00		25.00	
INDIANA				
Gary	0.00	Proportional cost of main extension.	25.00	Tap-in fee.
Indianapolis	Varies	Short tap.	50.00	
Fort Wayne	412.00		50.00	Permit only; wastewater area connection fee of $700 per acre.
	587.00	Long tap.		
IOWA				
Davenport	0.00		0.00	
Des Moines	Not applicable		0.00	
KANSAS				
Wichita	300.00		40.00	
KENTUCKY				
Louisville	425.00	Tapping and installation of individual water service.	0.00	

LOUISIANA				
Baton Rouge	74.00		0.00	
New Orleans	0.00		0.00	
Shreveport	Varies		Varies	
MARYLAND				
Baltimore	0.00		0.00	
MASSACHUSETTS				
Boston	235.00	Water service installation: $135. Water pipe inspection: $100.	200.00	Sewer pipe inspection: $100. Storm drain inspection: $100.
Salem	44.55	For a 5/8″ meter.	50.00	Sewer entry fee: residential & comm.
Springfield	54.25	For a 3/4″ meter.	200.00	Sewer entry fee: industrial
	8.25	Per foot (3/4″ piping) plus $65 meter charge (5/8″ meter).	0.00	
MICHIGAN				
Ann Arbor	1,005.00		420.00	
Detroit	0.00		0.00	
Flint	70.00		0.00	
Lansing	1,827.40		0.00	
Saginaw	Not applicable		Not applicable	
MINNESOTA				
Minneapolis	356.60	Inspection: $58.00 Street repair: $175.00 Permit and tap: $123.60	733.00	Inspection: $58.00 Street repair: $200.00 Sewer availability charge: $475.00
St. Paul	1,075.00	Service installation charge.	525.00	Sewer availability charge.
MISSISSIPPI				
Jackson	390.00		300.00	
MISSOURI				
Kansas City	Varies	Dependent on size of tap.	0.00	
St. Louis	55.00	Tapping fee.	350.00	

Total One-Time Charges Assessed for a New Single-Family Residence to Connect to the Water/Wastewater System

State/Cities	Water	Description	Wastewater	Description
NEBRASKA				
Omaha	$547.00		$50.00	Plus $20 per foot frontage charge. In some areas (where new development is) there is a $510 connection/interceptor sewer charge.
NEVADA				
Las Vegas	400.00		500.00	Connection fee per equivalent residential unit.
NEW JERSEY				
Jersey City	190.00	Tap fee: $100, Meter charge: $90.	Not applicable	
Newark	1,800.00	New connection charge.	Not applicable	
Trenton	0.00		5.00	
NEW MEXICO				
Albuquerque	1,001.00	Utility expansion charge only. Additional charge for service line installation.	524.00	Utility expansion charge only. Additional charge for service line installation
NEW YORK				
Albany	175.00	Meter charge (3/4"): $100 Tapping charge (3/4"): $75 Additional material & labor charge.	50.00	Tapping fee.
Buffalo	191.00		100.00	
New York City	57.00	Connection fee (3/4"–1"): $60 Meter fee (5/8"): $50	90.00	
Rochester	110.00		250.00	Per unit connection fee.
Syracuse	235.00	Paved, $0 unpaved. Meter charge: $50—5/8", $110—3/4"; $140—1"	0.00	

NORTH CAROLINA

City			
Charlotte	847.00	Connection: $475 Tapping privilege fee: $365 Deposit: $7	1,740.00	Connection: $1,000 Tapping privilege fee: $740
Greensboro	595.00	Plus $8 per front foot Note: If outside city limits, there is a $200 per acre charge for water and/or sewer.	340.00	Plus $9 per front foot.

OHIO

City			
Akron	780.00	Service tap.	265.00	Connection charge.
Canton	250.00		10.00	Inspection fee.
Cincinnati	1,500.00		80.00	Tap inspection fee plus $0–$5,000 per tap to recoup costs of sewer facilities serving the area of the tap.
Cleveland	185.00		0.00	
Columbus	1,645.00	Tap charge: $600 Meter charge: $88 System capacity charge: $357 Front foot charge (60'): $600	625.00	System capacity charge: $600 Sewer permit: $25
Dayton	1,300.00	Note: Both costs vary greatly depending on type of street to be cut and replaced, location, and depth of utility.	2,000.00	
Toledo	600.00		150.00	
Youngstown	485.00	Tap-in and meter.	600.00	

OKLAHOMA

City			
Oklahoma City	146.94	Impact charge: $100.00 Meter charge: $41.94 Occupancy certificate: $5.00	100.00	
Tulsa	175.00	Additional $300–$1,000 charged to builder for tap fee.	0.00	

OREGON

City			
Portland	710.00	Plus $150 if paving required.	885.00	Typical charge; $3,625 maximum.

Total One-Time Charges Assessed for a New Single-Family Residence to Connect to the Water/Wastewater System

State/Cities	Water	Description	Wastewater	Description
PENNSYLVANIA				
Allentown	$50.00	For 1″ tap.	$25.00	Average cost for permit to open street, which varies with size of opening.
Lancaster	0.00		250.00	
Philadephia	105.00	For ⁵⁄₈″ meter.	0.00	
Pittsburgh	188.00	Tap charge: $81 Meter charge (⁵⁄₈″): $107	100.00	
Scranton	0.00		25.00	(Soon to increase to $250).
York	Not available		0.00	
RHODE ISLAND				
Providence	765.90	Includes meter & service connection from main to curb, but doesn't include line from curb to house or any street repairs.	50.00 100.00	Residential connection fee. Non-residential connection fee.
SOUTH CAROLINA				
Charleston	865.00	Tap fee (³⁄₄″): $350 Impact fee (developer): $490	810.00	Tap fee (6″): $200 Impact fee: $610
Columbia	125.00		300.00	
Greenville	0.00		Not applicable	Not charged by authority.
TENNESSEE				
Chattanooga	0.00		500.00	
Johnson City	240.00		760.00	
Knoxville	400.00		675.00	Or $3.25 per month for 360 months sewer improvement charge.
Memphis	125.00		300.00	
Nashville	Varies		Varies	

TEXAS			
Austin	1,516.00	5/8" meter; 1,750 square feet of living area.	1,783.00	5/8" meter; 1,750 square feet of living area.
Beaumont	175.00		350.00	
Corpus Christi	Varies		Not available	
Dallas	175.00	Plus $40 deposit for residential.	100.00	
El Paso	215.00	Plus frontage fee of $6.90 per foot.	0.00	Plus frontage fee of $6.75 per foot.
Fort Worth	Varies		Varies	
Houston	Varies		Varies	
San Antonio	Varies		Not applicable	
UTAH				
Salt Lake City	230.00		1,000.00	
VIRGINIA				
Norfolk	350.00		652.00	Connection fee (existing): $250 Advance charge: $12 Tap charge (5/8"): $390
Richmond	865.00		800.00	
WASHINGTON				
Seattle	0.00		0.00	
Spokane	Actual cost		65.00	Connection fee; balance paid by private contractor.
Tacoma	Not available		Not available	
WISCONSIN				
Milwaukee	245.00		20.00	

C

MONTHLY WATER
CHARGES

Source: Arthur Young's 1988 National Water and Wastewater Rate Survey.

1988 National Water/Wastewater Rate Survey: Monthly Water Charges

| State/Cities | 5/8-Inch Meter | | | | 2-Inch Meter | 4-Inch Meter | 8-Inch Meter |
	0 Cubic Feet (0 Gallons)	500 Cubic Feet (3,740 Gallons)	1,000 Cubic Feet (7,480 Gallons)	3,000 Cubic Feet (22,440 Gallons)	50,000 Cubic Feet (374,000 Gallons)	1,000,000 Cubic Feet (7,480,000 Gallons)	1,500,000 Cubic Feet (11,220,000 Gallons)
ALABAMA							
Birmingham	$2.34	$6.34	$10.34	$26.13	$376.13	$7,063.60	$10,643.95
Mobile	3.12	3.89	7.73	22.87	344.22	4,529.16	6,207.18
ARIZONA							
Phoenix Summer:	3.92	5.67	7.42	21.17	371.20	7,149.55	Not applicable
Winter:	3.92	5.67	7.42	18.42	323.55	6,151.90	Not applicable
Tucson Summer:	3.80	8.20	13.35	46.05	70.55	Not applicable	Not applicable
Winter:	3.80	8.20	13.25	40.25	63.15	10,233.00	15,386.00
ARKANSAS							
Little Rock	3.60	5.82	9.52	24.32	317.87	3,168.15	4,684.20
CALIFORNIA							
Anaheim	7.95	9.94	11.93	19.89	226.70	4,055.00	6,200.00
Bakersfield	4.75	6.44	8.46	16.54	216.77	4,076.57	6,121.77
Fresno	2.25	3.10	3.95	7.35	91.10	1,714.20	2,584.45
Los Angeles Summer:	4.77	6.70	11.23	29.37	464.30	9,112.30	13,754.00
Winter:	4.77	6.18	10.20	26.28	412.80	8,082.30	12,209.00
Oakland	3.75	7.30	10.84	25.02	375.95	7,161.40	10,870.60
Sacramento	4.07	4.07	4.07	8.10	127.86	2,049.76	3,070.88
San Diego	3.12	7.47	11.81	31.93	486.34	9,398.74	14,321.22
San Francisco	1.50	4.05	6.60	16.80	256.50	5,101.50	7,650.00
San Jose	4.35	8.13	12.60	30.48	458.81	8,973.31	13,490.31
Stockton	4.55	5.82	7.08	12.14	130.25	2,116.90	3,197.90
Ventura	1.37	3.06	6.48	22.36	395.54	7,938.54	11,908.54

Note: A 1.5% California Public Utilities Commission reimbursement fee is added to all charges shown at billing date.

Note: An elevation charge is added to base rates which recovers the incremental cost of serving customers at elevations which require additional pumping.

Note: Water information from San Jose Water Company.

COLORADO							
Colorado Springs	2.74	9.67	16.59	44.30	695.39	13,855.74	20,782.24
Denver	2.15	5.25	8.36	19.58	207.64	3,517.00	5,357.80
CONNECTICUT							
Hartford	4.85	9.20	13.55	30.95	443.00	6,654.40	9,853.40
New Haven	6.52	13.39	20.25	47.71	602.95	10,425.56	15,429.73
DISTRICT OF COLUMBIA							
Washington	0.00	5.04	10.04	30.12	502.00	10,040.00	15,060.00
FLORIDA							
Fort Lauderdale	2.46	6.13	9.79	24.45	382.05	7,377.99	11,146.28
Jacksonville	5.54	5.54	9.40	17.00	211.00	2,847.00	4,336.00
Lakeland	3.10	4.80	7.35	20.10	306.67	5,880.10	8,902.40
Miami	3.96	3.96	6.58	19.74	328.95	6,579.41	9,869.11
Orlando	2.15	3.73	5.85	13.77	177.04	3,350.16	5,036.31
St. Petersburg	2.20	5.98	9.75	25.51	388.34	7,585.60	11,428.65
Tampa	1.50	3.05	6.10	18.30	305.00	6,100.00	9,150.00
West Palm Beach	4.20	6.40	9.10	38.20	276.60	Not applicable	Not applicable
GEORGIA							
Atlanta	3.35	6.75	15.25	49.25	564.45	7,458.85	11,058.85
Augusta	2.40	2.40	5.60	17.60	251.04	3,519.80	5,090.60
HAWAII							
Honolulu	1.25	4.19	7.55	20.15	316.66	6,285.70	9,427.30
ILLINOIS							
Chicago Gross:	0.00	2.96	5.91	17.73	295.50	5,910.00	8,865.00
Net:	0.00	2.89	5.77	17.31	288.50	5,770.00	8,655.00
Joliet	2.55	6.96	14.31	41.51	586.71	11,606.71	17,406.71
Peoria	5.00	13.15	21.30	53.90	427.03	6,411.53	9,667.53

Note: Net rate if paid within 21 days of biling date.

State/Cities	0 Cubic Feet (0 Gallons)	5/8-Inch Meter			2-Inch Meter 50,000 Cubic Feet (374,000 Gallons)	4-Inch Meter 1,000,000 Cubic Feet (7,480,000 Gallons)	8-Inch Meter 1,500,000 Cubic Feet (11,220,000 Gallons)
		500 Cubic Feet (3,740 Gallons)	1,000 Cubic Feet (7,480 Gallons)	3,000 Cubic Feet (22,440 Gallons)			
INDIANA							
Gary	$7.38	$8.88	$16.47	$45.67	$474.11	$6,187.89	$9,179.89
Indianapolis	3.25	8.05	12.85	30.55	404.30	4,469.75	6,440.50
Fort Wayne	3.59	7.13	10.67	24.83	78.24	628.74	968.03
IOWA							
Davenport	3.35	7.96	12.57	31.01	85.81	6,178.60	8,727.85
Des Moines	3.15	4.72	9.43	28.29	428.67	7,333.11	10,963.11
KANSAS							
Wichita	4.30	5.01	8.60	22.96	219.85	3,528.38	4,897.36
KENTUCKY							
Louisville	3.15	6.49	10.37	26.98	427.08	7,527.82	11,370.27
LOUISIANA							
Baton Rouge	7.23	8.98	13.37	30.91	344.40	4,147.20	5,867.20
New Orleans	2.80	9.57	16.53	40.54	628.16	9,847.60	14,630.00
Shreveport	2.10	5.80	9.50	24.32	306.50	5,263.10	7,933.25
MARYLAND							
Baltimore	1.98	3.95	5.93	14.74	147.82	2,522.82	3,772.82
MASSACHUSETTS							
Boston	0.00	4.41	8.82	26.52	445.97	8,983.07	13,478.07
Salem	10.50	10.50	10.50	31.50	525.00	10,500.00	15,750.00
Springfield	2.33	4.83	7.33	16.53	168.58	2,289.83	3,479.50
MICHIGAN							
Ann Arbor	2.33	4.55	9.10	27.30	455.00	9,100.00	13,650.00

Note: Charges are subject to a 10% discount if paid within 30 days.

Detroit	0.80	2.79	4.77	12.71	175.89	3,077.49	4,641.49
Flint	2.54	6.99	11.44	29.24	446.50	8,865.89	10,763.07
Lansing	4.15	8.50	12.85	30.25	474.43	8,866.00	13,631.00
Saginaw	1.83	3.19	5.01	11.82	190.47	3,211.92	4,952.88
MINNESOTA							
Minneapolis	1.00	4.25	8.50	25.50	425.00	8,500.00	12,750.00
St. Paul	3.20	7.55	11.90	26.10	445.10	8,456.50	12,732.40
MISSISSIPPI							
Jackson	0.00	5.00	10.00	30.00	500.00	10,000.00	15,000.00

Note: Customers pay a monthly "meter charge" for capital improvements to water system (5/8" – $2.50, 1" – $6.25, 1 1/2-2" – $20.00, over 2" – $35.00).

MISSOURI							
Kansas City	4.50	8.40	12.30	27.90	321.60	5,944.30	8,404.00
St. Louis	2.79	5.27	8.99	21.39	386.27	4,962.32	7,469.23
NEBRASKA							
Omaha Summer:	1.85	4.18	6.86	17.96	244.80	3,994.70	5,919.00
Winter:	1.85	4.18	6.50	15.80	209.30	3,994.70	5,919.00
NEVADA							
Las Vegas	8.66	11.39	14.12	25.04	323.93	5,614.11	8,676.72
NEW JERSEY							
Jersey City	1.00	4.75	8.50	23.50	383.00	7,530.00	11,340.00

Note: Above rates do not include fire protection charge of $0.533 per day for 2" meter, $2 per day for 4" meter, and $6 per day for 8" meter.

Newark	10.37	10.37	15.56	36.30	484.09	8,042.09	11,767.09
Trenton	4.48	5.49	6.50	10.56	145.44	2,076.46	3,597.40
NEW MEXICO							
Albuquerque	4.00	6.35	8.70	18.10	260.89	4,725.89	8,198.42
NEW YORK							
Albany	3.75	8.75	13.75	33.75	503.75	11,059.75	16,659.75
Buffalo	6.36	6.36	6.36	19.08	190.96	3,344.96	5,004.96
New York City	4.05	4.05	8.10	24.30	405.00	8,100.00	12,150.00

State/Cities	0 Cubic Feet (0 Gallons)	500 Cubic Feet (3,740 Gallons)	5/8-Inch Meter 1,000 Cubic Feet (7,480 Gallons)	3,000 Cubic Feet (22,440 Gallons)	2-Inch Meter 50,000 Cubic Feet (374,000 Gallons)	4-Inch Meter 1,000,000 Cubic Feet (7,480,000 Gallons)	8-Inch Meter 1,500,000 Cubic Feet (11,220,000 Gallons)
NEW YORK continued							
Rochester	$2.48	$7.64	$12.77	$33.08	$489.34	$7,898.50	$11,844.58
Syracuse	2.97	3.43	6.86	20.58	249.00	4,353.00	5,843.00
NORTH CAROLINA							
Charlotte	1.40	4.40	7.40	19.40	301.40	6,001.40	9,001.40
Greensboro	2.10	3.50	7.00	21.00	247.50	3,097.50	4,597.50
Raleigh	1.41	6.61	11.81	32.61	525.80	10,416.43	15,650.86
OHIO							
Akron	2.02	8.37	14.72	40.12	524.68	8,014.81	11,075.50
Canton	2.00	2.00	3.03	9.10	104.37	1,133.33	1,700.00
Cincinnati	2.65	3.83	6.78	17.68	256.03	4,460.23	6,700.73
Cleveland	1.73	2.86	6.23	18.68	335.46	6,738.46	10,108.46
Columbus	2.84	5.97	9.09	22.57	334.03	5,905.33	8,721.03
Dayton	3.66	3.66	3.66	8.79	140.33	2,170.37	3,156.73

Note: Rates shown would be gross rates. Net rates would be discounted 5% if paid within 14 days.

State/Cities	0 Cubic Feet (0 Gallons)	500 Cubic Feet (3,740 Gallons)	5/8-Inch Meter 1,000 Cubic Feet (7,480 Gallons)	3,000 Cubic Feet (22,440 Gallons)	2-Inch Meter 50,000 Cubic Feet (374,000 Gallons)	4-Inch Meter 1,000,000 Cubic Feet (7,480,000 Gallons)	8-Inch Meter 1,500,000 Cubic Feet (11,220,000 Gallons)
Toledo	6.05	6.05	6.05	18.15	295.30	4,821.80	6,703.40
Youngstown	0.00	1.71	3.02	8.20	101.69	1,364.52	2,046.78
OKLAHOMA							
Oklahoma City	2.70	4.74	8.82	24.12	383.16	7,631.28	11,446.08
Tulsa	3.60	3.82	7.63	22.89	347.83	6,956.44	10,434.71
OREGON							
Portland	2.25	5.30	8.35	20.55	311.80	6,117.15	9,195.40

PENNSYLVANIA						
Allentown	2.52	5.90	9.28	22.81	344.89	6,790.69	10,145.12
Lancaster	1.65	4.76	9.53	29.03	357.26	3,768.15	5,563.35
Philadelphia	2.08	6.81	11.53	28.28	377.08	6,477.97	9,708.30
Pittsburgh	4.20	8.27	14.58	42.11	639.36	12,622.50	18,985.48
Scranton	4.86	9.29	16.17	43.71	322.84	5,665.10	8,233.25
RHODE ISLAND						
Providence	1.19	2.74	4.29	10.49	137.50	1,769.87	2,656.37
SOUTH CAROLINA						
Charleston	2.93	5.18	8.23	20.43	241.36	3,828.39	5,758.82
Columbia	2.40	3.90	7.65	22.65	361.20	6,579.35	9,829.35
Greenville	2.35	3.29	6.58	18.92	193.10	3,116.80	4,612.80
TENNESSEE						
Chattanooga	6.11	7.73	15.89	48.49	602.09	7,212.14	10,652.58
Johnson City	4.39	9.62	16.77	42.19	530.46	8,652.18	12,915.78
Knoxville	6.25	9.53	17.73	50.53	603.33	7,211.33	10,373.33
Memphis	2.12	2.81	5.62	16.16	225.06	2,877.59	4,268.59
Nashville	3.13	8.95	18.65	57.45	704.61	10,407.78	15,416.15
TEXAS						
Austin	5.46	9.18	17.19	49.20	811.40	16,063.10	24,132.33
Beaumont	1.23	6.02	10.80	29.95	479.95	9,575.63	14,362.83
Corpus Christi	3.58	5.84	10.85	34.40	394.36	6,096.64	8,868.68
Dallas Summer:	1.29	4.92	9.35	18.29	314.70	5,270.20	7,976.72
Winter:	1.29	4.92	9.35	17.47	275.21	5,270.20	7,976.72
El Paso	3.13	3.59	5.89	17.13	416.84	8,497.43	12,780.18
Fort Worth Summer:	2.40	8.05	13.70	36.30	573.81	6,636.45	9,696.50
Winter:	2.40	8.05	13.70	36.30	408.20	6,470.85	9,530.90
Houston	4.47	10.40	15.98	58.58	1,235.39	24,912.01	37,403.41
San Antonio	5.09	6.65	9.26	25.31	538.88	10,860.83	16,313.19

State/Cities	5/8-Inch Meter				2-Inch Meter	4-Inch Meter	8-Inch Meter
	0 Cubic Feet (0 Gallons)	500 Cubic Feet (3,740 Gallons)	1,000 Cubic Feet (7,480 Gallons)	3,000 Cubic Feet (22,440 Gallons)	50,000 Cubic Feet (374,000 Gallons)	1,000,000 Cubic Feet (7,480,000 Gallons)	1,500,000 Cubic Feet (11,220,000 Gallons)
UTAH							
Salt Lake City	$5.42	$5.42	$5.42	$13.02	$209.87	$3,869.19	$5,928.97
VIRGINIA							
Norfolk	2.13	7.23	12.33	30.60	472.23	9,217.86	13,901.46
Richmond	5.83	9.72	13.60	29.14	318.41	4,912.36	7,449.34
WASHINGTON							
Seattle	1.40	4.48	7.55	12.69	178.40	3,453.40	5,217.00
Spokane	3.83	5.68	7.53	14.93	200.24	2,289.75	3,357.35
Tacoma	5.63	7.55	9.41	15.67	165.32	2,539.50	4,028.80
WISCONSIN							
Milwaukee	1.87	4.57	7.27	18.07	276.53	4,407.20	6,206.20
AVERAGE MONTHLY BILL:	$3.32	$6.12	$9.95	$26.09	$355.47	$6,275.29	$9,507.20

D   MONTHLY WASTEWATER CHARGES

Source: Arthur Young's 1988 National Water and Wastewater Rate Survey.

1988 National Water/Wastewater Rate Survey: Monthly Wastewater Charges

State/Cities	0 Cubic Feet (0 Gallons)	5/8-Inch Meter 500 Cubic Feet (3,740 Gallons)	5/8-Inch Meter 1,000 Cubic Feet (7,480 Gallons)	5/8-Inch Meter 3,000 Cubic Feet (22,440 Gallons)	2-Inch Meter 50,000 Cubic Feet (374,000 Gallons)	4-Inch Meter 1,000,000 Cubic Feet (7,480,000 Gallons)	8-Inch Meter 1,500,000 Cubic Feet (11,220,000 Gallons)
ALABAMA							
Birmingham	$1.20	$5.86	$10.76	$30.36	$498.26	$9,826.26	$14,804.76
Mobile	5.22	6.49	12.88	37.89	558.28	8,257.96	11,600.25
ARIZONA							
Phoenix	4.00	4.00	4.87	13.48	233.56	4,592.16	Not applicable
Tucson	1.74	4.19	7.59	21.12	339.90	6,782.63	10,173.59
ARKANSAS							
Little Rock	3.65	5.84	9.49	24.09	401.79	7,425.94	11,202.34
CALIFORNIA							
Bakersfield	6.50	6.50	6.50	11.20	185.39	3,709.97	5,565.00
Fresno	4.37	4.37	4.37	4.37	Varies	Varies	Varies
Los Angeles							
Residential:	Not applicable	1.66	3.32	9.96	166.00	3,330.00	4,980.00
Commercial:	Not applicable	2.56	5.11	15.33	255.50	5,110.00	7,665.00
Oakland	4.42	5.31	6.19	6.19	607.35	12,102.35	18,152.35
Sacramento							
Collection fee:	3.10	3.10	3.10	3.10	59.78	813.24	1,219.86
Regional treatment fee:	7.30	7.30	7.30	7.30	226.30	4,547.90	6,825.50
San Diego	10.40	10.40	10.40	10.40	369.00	7,380.00	11,070.00
San Francisco							
Residential:	0.00	5.85	14.17	47.45	829.40	16,634.94	24,953.64
Commercial:	0.00	8.83	17.65	52.96	882.70	17,653.90	26,480.85

Note: Bakersfield rate for use greater than 3,00 cubic feet does not include charges for BOD and SS.

San Jose	12.20	12.20	12.20	12.20	Varies	Varies	Varies

Note: Information from City of San Jose. Charges are based on volume and type of user.

Stockton	9.75	9.75	9.75	9.75	320.10	6,305.10	9,455.10
Ventura	5.22	5.22	8.36	11.13	402.69	7,993.19	11,988.20
COLORADO							
Colorado Springs	6.90	9.35	11.80	21.60	251.90	4,906.90	7,356.90
Denver	2.60	3.55	7.10	21.32	355.30	7,106.00	10,659.00
CONNECTICUT							
Hartford	Not applicable	Not applicable	Not applicable	Not applicable	435.00	8,700.00	13,050.00

Note: Adjusted ad valorem system — residential costs are recovered through property taxes.

New Haven	5.72	5.72	5.72	10.29	152.82	3,034.52	4,551.19
DISTRICT OF COLUMBIA							
Washington	0.00	9.32	18.64	55.92	932.00	18,640.00	27,960.00
FLORIDA							
Fort Lauderdale Residential:	1.46	7.18	12.90	32.06	42.00	66.16	144.61
Commercial:	1.46	7.18	12.90	35.79	583.62	11,479.96	17,280.61
Jacksonville	8.80	8.80	26.30	76.30	1,251.30	25,001.30	37,501.30
Lakeland	5.15	9.15	13.15	28.15	435.80	7,711.75	11,735.00
Miami	1.91	5.11	8.29	21.06	321.04	6,384.59	9,575.94
Orlando	8.44	13.44	18.44	23.44	918.20	18,369.00	27,552.60
St. Petersburg	3.70	9.09	14.47	32.50	560.36	10,836.85	16,364.20
Tampa	0.00	6.20	12.40	37.20	620.00	12,400.00	18,600.00
West Palm Beach	6.90	10.30	14.20	18.10	446.60	Not applicable	Not applicable
GEORGIA							
Atlanta	0.00	6.00	12.00	36.00	600.00	12,000.00	18,000.00
Augusta	2.60	4.04	5.12	10.52	115.57	1,586.51	2,293.37

State/Cities		5/8-Inch Meter				2-Inch Meter	4-Inch Meter	8-Inch Meter
		0 Cubic Feet (0 Gallons)	500 Cubic Feet (3,740 Gallons)	1,000 Cubic Feet (7,480 Gallons)	3,000 Cubic Feet (22,440 Gallons)	50,000 Cubic Feet (374,000 Gallons)	1,000,000 Cubic Feet (7,480,000 Gallons)	1,500,000 Cubic Feet (11,220,000 Gallons)
HAWAII								
Honolulu		$10.25	$10.25	$10.25	$17.05	$284.24	$5,684.80	$8,527.20
ILLINOIS								
Chicago	Gross:	0.00	1.42	2.84	8.51	141.84	2,836.80	4,255.20
	Net:	0.00	1.39	2.77	8.31	138.48	2,769.60	4,154.40
				Note: Net rate if paid within 21 days of biling date.				
Joliet		2.51	7.10	14.75	45.35	764.45	15,399.45	23,049.45
Peoria		0.00	2.74	5.48	17.46	291.00	6,301.00	9,451.50
INDIANA								
Gary		0.00	4.68	9.35	28.05	467.50	9,350.00	Not applicable
Indianapolis		5.43	6.28	10.53	27.54	427.24	8,506.28	12,758.40
Industrial:		5.59	6.48	10.94	28.75	447.45	8,910.53	13,364.78
Fort Wayne		2.89	2.99	5.98	17.94	29.90	597.90	896.85
IOWA								
Des Moines		4.00	5.01	10.02	30.07	501.16	10,023.20	15,034.80
KANSAS								
Wichita		3.69	4.07	6.02	13.80	196.61	3,891.73	5,836.53
KENTUCKY								
Louisville		4.70	8.33	11.95	26.47	379.03	7,309.75	11,039.84
LOUISIANA								
Baton Rouge		5.64	6.65	11.77	32.23	513.17	10,234.18	15,350.50
New Orleans		5.60	9.51	13.66	29.54	425.05	8,107.08	12,296.62
Shreveport		0.97	5.36	9.76	27.34	364.39	7,257.21	10,885.01

MARYLAND							
Baltimore	3.31	6.61	9.92	24.65	283.88	5,299.88	7,939.88
MASSACHUSETTS							
Boston	0.00	5.05	10.15	30.80	535.08	11,051.88	16,596.88
Salem	0.00	6.45	12.90	38.70	765.00	18,800.00	28,200.00
MICHIGAN							
Ann Arbor	4.26	8.10	16.20	48.60	810.00	16,200.00	24,300.00
Detroit	2.87	5.66	8.45	19.61	281.87	5,591.81	8,381.81
Flint	5.88	10.38	14.88	32.88	480.83	9,282.59	14,119.59
Lansing	0.74	6.74	12.74	36.74	600.74	12,000.74	18,000.74
Saginaw	2.00	3.99	6.65	16.61	270.97	5,053.30	7,787.29
MINNESOTA							
Minneapolis	1.67	6.00	12.00	36.00	600.00	11,730.00	17,430.00
St. Paul	3.60	7.30	14.60	43.80	730.00	14,060.00	20,760.00
MISSISSIPPI							
Jackson	2.81	5.20	10.40	31.20	520.00	10,400.00	15,600.00
MISSOURI							
Kansas City	2.10	4.10	6.10	14.10	202.40	4,002.40	6,002.40
St. Louis	4.31	4.31	4.31	13.45	154.45	3,004.45	4,504.45
NEBRASKA							
Omaha	4.20	6.55	8.70	17.30	221.00	4,543.15	6,693.15
NEVADA							
Las Vegas	6.83	6.83	6.83	6.83	341.67	6,812.83	10,222.67
NEW JERSEY							
Jersey City	0.00	7.50	15.00	45.00	750.00	15,000.00	22,500.00
Newark	13.40	13.40	13.40	20.10	335.00	6,700.00	10,050.00
Trenton	0.83	5.63	10.43	29.43	480.83	9,600.83	14,400.83

Note: Charges are subject to a 10% discount if paid within 30 days.

State/Cities	0 Cubic Feet (0 Gallons)	5/8-Inch Meter 500 Cubic Feet (3,740 Gallons)	1,000 Cubic Feet (7,480 Gallons)	3,000 Cubic Feet (22,440 Gallons)	2-Inch Meter 50,000 Cubic Feet (374,000 Gallons)	4-Inch Meter 1,000,000 Cubic Feet (7,480,000 Gallons)	8-Inch Meter 1,500,000 Cubic Feet (11,220,000 Gallons)
NEW MEXICO							
Albuquerque	$4.79	$7.09	$9.39	$18.59	$310.46	$4,793.88	$8,549.93
NEW YORK							
Albany	3.00	7.00	11.00	27.00	403.00	8,847.80	13,327.80
Buffalo	7.33	7.33	7.33	15.15	252.50	5,050.00	7,575.00
New York City	2.84	2.84	5.67	17.01	283.50	5,670.00	8,505.00
Rochester	0.00	3.04	6.08	18.23	303.75	6,075.00	9,112.50
Note: Operation and maintenance costs only. Capital rate is $0.91535 per $1,000 assessed value.							
Syracuse	0.00	1.12	2.24	6.72	111.50	2,230.00	3,345.00
NORTH CAROLINA							
Charlotte	1.40	5.65	9.90	26.90	426.40	8,501.40	12,571.40
Greensboro	1.89	3.15	6.30	18.90	315.00	6,300.00	9,450.00
Raleigh	0.84	4.54	8.24	23.04	370.84	7,400.84	11,100.84
OHIO							
Akron	1.09	6.78	12.46	35.20	543.09	12,801.09	19,201.09
Canton	2.20	2.20	3.00	8.33	401.00	2,667.00	4,000.33
Cincinnati	6.50	8.26	12.66	30.26	408.36	7,718.60	11,738.60
Cleveland	2.85	4.28	8.55	25.65	427.50	8,550.00	12,825.00
Columbus	1.42	5.75	10.07	27.37	433.92	8,657.46	12,982.46
Dayton	4.42	4.42	4.61	8.48	93.11	1,614.64	2,415.22
Note: Rates shown are gross rates. Net rates would be discounted 5% if paid within 14 days							
Toledo	5.59	10.00	14.41	32.05	456.33	8,850.68	13,336.07
Youngstown	0.00	5.42	6.58	11.25	120.92	2,333.34	3,500.00
OKLAHOMA							
Oklahoma City	0.00	6.00	14.00	44.00	748.00	14,960.00	22,440.00
Tulsa	3.19	4.90	9.80	29.40	489.95	9,798.82	14,698.25

OREGON						
Portland	6.15	9.70	13.24	27.42	360.65	7,096.15	10,641.15
PENNSYLVANIA						
Allentown	1.20	4.65	8.10	21.89	349.77	6,911.29	10,388.96
Lancaster	2.12	5.35	10.71	32.61	436.93	6,832.33	10,198.33
Philadelphia	11.77	14.53	17.28	28.30	338.83	5,773.33	9,051.67
Pittsburgh	0.65	1.28	2.25	6.50	98.34	1,947.52	2,920.00

Note: Sewage collection only. Does not include treatment cost.

Scranton	5.61	7.48	13.09	41.14	669.53	10,899.85	15,836.65
York	6.37	10.96	19.39	47.96	495.08	7,108.70	10,549.50
RHODE ISLAND						
Providence						
Non-residential:	0.00	7.00	14.00	42.00	700.00	14,000.00	21,000.00

Residential Rate: $62.50 per dwelling unit per year, billed semi-annually.

SOUTH CAROLINA						
Charleston	5.64	10.42	19.98	63.00	1,023.17	15,413.17	22,863.17
Columbia	1.52	5.27	9.02	24.02	376.52	7,501.52	11,251.52
Greenville	4.54	10.42	16.43	40.22	569.28	11,299.34	16,946.74
TENNESSEE						
Chattanooga	4.88	9.46	18.92	56.77	762.64	9,133.30	13,097.70
Johnson City	4.39	9.62	16.77	42.19	530.46	8,652.18	12,915.78
Knoxville	5.41	7.99	14.44	40.24	562.14	10,159.14	15,209.14
Memphis	1.50	2.19	4.39	13.17	219.48	4,389.60	6,584.40
Nashville	3.13	8.95	18.65	57.45	704.61	10,407.78	15,416.15

Note: Sprinkler allowances for months of June, July, and August.

TEXAS						
Austin	5.61	11.84	25.23	78.79	1,337.37	26,776.85	40,166.06
Beaumont	1.23	5.79	10.36	28.61	457.51	9,126.83	13,689.63
Corpus Christi	5.50	7.03	10.32	16.94	335.61	6,588.89	9,880.09

State/Cities	5/8-Inch Meter				2-Inch Meter	4-Inch Meter	8-Inch Meter
	0 Cubic Feet (0 Gallons)	500 Cubic Feet (3,740 Gallons)	1,000 Cubic Feet (7,480 Gallons)	3,000 Cubic Feet (22,440 Gallons)	50,000 Cubic Feet (374,000 Gallons)	1,000,000 Cubic Feet (7,480,000 Gallons)	1,500,000 Cubic Feet (11,220,000 Gallons)
TEXAS continued							
Dallas	$1.65	$9.47	$17.28	$29.92	$472.89	$9,426.45	$14,138.85
El Paso	4.70	5.21	7.76	10.00	260.14	5,107.62	7,662.58
Fort Worth	4.30	7.50	10.70	23.45	323.80	6,394.30	9,589.30
Houston	3.56	8.28	15.36	50.76	881.48	17,651.64	26,478.04
San Antonio	5.40	8.28	13.03	32.54	469.64	9,304.64	13,954.64
UTAH							
Salt Lake City	7.10	7.10	7.10	7.10	276.00	5,520.00	8,280.00
VIRGINIA							
Norfolk	2.75	4.39	8.49	24.89	383.89	7,603.89	11,403.89
			Note: Costs above are for treatment only.				
Richmond	7.84	12.58	17.32	36.28	552.28	9,801.00	14,541.00
WASHINGTON							
Seattle	1.17	10.27	19.37	55.77	911.17	18,201.17	27,301.17
Spokane	8.85	8.85	8.85	8.85	195.94	3,801.19	5,698.69
Tacoma	5.23	9.53	13.83	31.03	582.18	11,576.53	17,363.03
	Note: Above rates for 2", 4", and 8" meters include only minimum high-strength surcharges.						
WISCONSIN							
Milwaukee	3.42	5.72	8.02	17.22	240.05	4,630.49	6,976.63
AVERAGE MONTHLY BILL:	$3.78	$6.71	$10.84	$26.84	$441.80	$8,502.24	$2,784.12

E EFFECTIVE DATES, RATE STRUCTURE AND BILLING CYCLE

Source: Arthur Young's 1988 National Water and Wastewater Rate Survey.

1988 National Water/Wastewater Rate Survey: Effective Dates, Rate Structure and Billing Cycle

	Effective Dates		Rate Structure (Number of Blocks)		Billing Cycle	
	Water	Wastewater	Water	Wastewater	Water	Wastewater
ALABAMA						
Birmingham	Mar 87	Nov 83	DB(5)	IB(2)	Monthly	Monthly
Mobile	Jan 87	Jan 87	DB(9)	DB(9)	Monthly	Monthly
ARIZONA						
Phoenix	Apr 87	Jan 87	IB(3)	Varies with type of user	Monthly	Monthly
Tucson	May 87	Feb 87	IB(7)	U	Monthly	Monthly
ARKANSAS						
Little Rock	Feb 85	Feb 87	DB(5)	U	Monthly	Monthly
CALIFORNIA						
Anaheim	Aug 87	Not available	U	Not available	Residential: bimonthly Commercial: monthly	Not available
Bakersfield	Mar 87	Jul 87	IB(2) Residential customers have option of service under flat rate structure	U	Monthly	Flat rate: annually Surcharges: semiannually
Fresno	Sep 86	Apr 87	U	U	Bimonthly	Bimonthly
Los Angeles	Nov 87	Aug 87	U	U	Monthly & bimonthly	Monthly & bimonthly
Oakland	Jul 87	Jul 87	U	U	Bimonthly (Note: 500 largest accounts billed monthly)	Bimonthly
Sacramento	Jul 87	Jul 87	U and DB(3) (U – Residential, DB – Commercial)	U	Bimonthly	Monthly
San Diego	Jul 87	Jul 87	U and IB(2) (IB – Residential, U – Commercial)	U and IB(11) (U – Residential, IB – Commercial)	Bimonthly	Bimonthly

City	Effective Date	Effective Date	Rate Structure	Rate Structure	Billing Cycle	Billing Cycle
San Francisco	Jul 86	Aug 84	U	U and IB(2) (IB – Residential, U – Commercial)	Large: monthly / Small: bimonthly	Large: monthly / Small: bimonthly
San Jose	Jul 87	Jul 87	IB(2)	U	Monthly	Heavy: monthly / All other: annually
Stockton	Jul 87	Jul 87	DB(2)	U	Monthly	Quarterly
Ventura	Jul 87	Jun 87	IB(2)	IB (3) – Residential (6) – Commercial	Bimonthly	Bimonthly
COLORADO						
Colorado Springs	Jan 87	Jan 86	U	U	Monthly	Monthly
Denver	Mar 80	Apr 87	DB(4)	U	Bimonthly	Monthly & bimonthly
CONNECTICUT						
Hartford	Jan 87	Mar 87	DB(2)	U	Monthly & quarterly	Monthly & quarterly
New Haven	Jul 87	Aug 86	DB(3)	U	Quarterly	Monthly & quarterly
DISTRICT OF COLUMBIA						
Washington	Oct 87	Oct 87	U	U	Quarterly (some monthly)	Quarterly (some monthly)
FLORIDA						
Fort Lauderdale	Oct 87	Oct 87	U	U	Bimonthly	Bimonthly
Jacksonville	Dec 81	Dec 81	IB(2)	IB(2)	Monthly	Monthly
Lakeland	Oct 84	Oct 84	DB(3)	U	Monthly	Monthly
Miami	Oct 86	Oct 86	U	U	Quarterly	Quarterly
Orlando	Jan 87	Feb 87	DB(2)	U	Monthly	Monthly
St. Petersburg	Sep 87	Sep 87	IB(3) – Residential / U – Commercial	U	Not available	Not available
Tampa	Oct 87	Oct 83	U	U	Monthly	Monthly
West Palm Beach	Oct 87	Oct 87	IB(2) – Residential / U – Commercial	IB(2) – Residential / U – Commercial	Monthly	Monthly

LEGEND: DB = Declining Block; IB = Inverted Block; U = Uniform rate structure; (#) = Number of rate blocks.

	Effective Dates		Rate Structure (Number of Blocks)		Billing Cycle	
	Water	Wastewater	Water	Wastewater	Water	Wastewater
GEORGIA						
Atlanta	Mar 84	Mar 84	DB(4)	U	Residential: bimonthly Commercial: monthly	Residential: bimonthly Commercial: monthly
Augusta	Jan 80	Jan 80	DB(5)	DB(5)	Monthly	Monthly
HAWAII						
Honolulu	Aug 85	Jul 84	U	U	Bimonthly; monthly for large accounts	Bimonthly; monthly for large accounts
ILLINOIS						
Chicago	May 85	Jan 84	U	U	Assessed: semiannually Metered: bimonthly Large users: monthly	Assessed: semiannually Metered: bimonthly Large users: monthly
Joliet	Apr 85	Jul 86	DB(3)	IB(2)	Monthly	Monthly
Peoria	Mar 86	Jun 87	DB(4)	U	Monthly & quarterly	Monthly & quarterly
INDIANA						
Gary	Aug 87	Jan 86	DB(6)	U	Residential: bimonthly Commercial: monthly	Bimonthly
Indianapolis	Jul 86	Jan 85	DB(5)	U	Residential: bimonthly Commercial: monthly	Monthly
Fort Wayne	Aug 86	Apr 86	DB(3)	U	Residential: bimonthly Commercial: monthly	Monthly
IOWA						
Davenport	Jul 87	Jul 87	DB(4)	U	Quarterly	Residential: quarterly Commercial: quarterly Industrial: monthly
Des Moines	Jan 83	Feb 87	DB(3)	U	Monthly	Monthly
KANSAS						
Wichita	Jan 87	Jan 87	DB	U	Bimonthly	Bimonthly

KENTUCKY						
Louisville	Jan 87	Jul 86	IB(6)	U	Monthly & bimonthly	Monthly & bimonthly
LOUISIANA						
Baton Rouge	Jun 87	Jul 87	DB(5)	U	Monthly	Monthly
New Orleans	Jan 87	Jan 86	DB(3)	U	Monthly	Monthly
Shreveport	Jan 85	Jan 85	U	U	Monthly	Monthly
MASSACHUSETTS						
Boston	Apr 87	Apr 87	IB(10)	IB(10)	Quarterly	Quarterly
Salem	1987	1987	U	U	Quarterly	Quarterly
Springfield	City—1978 Suburbs—1986	Not available	DB(3)	Not available	Semiannually, quarterly	Not available
MICHIGAN						
Ann Arbor	Jul 85	Jul 87	U	U	Quarterly (a few monthly)	Quarterly (a few monthly)
Detroit	Jul 87	Jul 87	DB(3)	U	Residential: quarterly / Commercial: monthly	Residential: quarterly / Commercial: monthly
Flint	Jul 87	Jul 87	DB(3)	U	Monthly	Monthly
Lansing	Nov 86	Jul 87	U	U	Residential: quarterly / General Svc. quarterly / Industrial: monthly	Residential: quarterly / General Svc. quarterly / Industrial: monthly
Saginaw	May 85	May 85	DB(3)	U	Residential: quarterly / Wholesale: monthly / Industrial: monthly	Residential: quarterly / Wholesale: monthly / Industrial: monthly
MINNESOTA						
Minneapolis	Jan 84	Feb 87	U	DB(5)	Quarterly (some monthly)	Quarterly (some monthly)
St. Paul	Jan 87	Jan 87	DB(3)	DB(3)	Residential: quarterly / Commercial: monthly	Residential: quarterly / Commercial: monthly

LEGEND: DB = Declining Block; IB = Inverted Block; U = Uniform rate structure; (#) = Number of rate blocks.

	Effective Dates		Rate Structure (Number of Blocks)		Billing Cycle	
	Water	Wastewater	Water	Wastewater	Water	Wastewater
MISSISSIPPI						
Jackson	Jun 86	Nov 86	U	U	Bimonthly	Bimonthly
MISSOURI						
Kansas City	May 87	May 87	DB(3)	U	Residential: bimonthly All others: monthly	Residential: bimonthly All others: monthly
St. Louis	Sep 86	Jul 86	DB(3)	U	Quarterly	Semiannually
NEBRASKA						
Omaha	Mar 87	Jan 87	IB(3)	IB(2)	Monthly	Monthly
NEVADA						
Las Vegas	Oct 87	Jul 86	U	U	Monthly	Monthly
NEW JERSEY						
Jersey City	Jan 82	Jan 87	U	U	Quarterly	Residential: quarterly Industrial: monthly
Newark	Feb 84	Feb 84	DB(5)	U	Quarterly	Quarterly
Trenton	Mar 84	Mar 83	DB(3)	U	Quarterly	Quarterly
NEW MEXICO						
Albuquerque	Aug 82	Aug 82	U	U	Monthly	Monthly
NEW YORK						
Albany	Nov 83	Nov 83	IB(2)	IB(2)	Triannually	Triannually
Buffalo	Jul 82	Jun 82	DB(3)	U	Monthly & quarterly	Monthly & quarterly
New York City	Jul 87	Jul 87	U	U	Meter size: Large: bimonthly Small: semiannually Meters > 3″: monthly All others: quarterly	Meter size: Large: bimonthly Small: semiannually Quarterly, monthly & annually
Rochester	Jun 87	Jan 87	DB(3)	U	Residential: semiannual All others: quarterly Industrial: monthly	Residential: semiannual All others: quarterly Industrial: monthly
Syracuse	Dec 84	Dec 84	DB(4)	U		

NORTH CAROLINA						
Charlotte	Jul 87	Jul 87	U	U	Monthly	Monthly
Greensboro	Jul 86	Jul 86	DB(3)	DB(3)	Monthly & quarterly	Monthly & quarterly
Raleigh	Aug 86	Aug 86	U	U	Monthly	Monthly
OHIO						
Akron	Dec 86	Apr 87	DB(3)	DB(3)	Monthly	Monthly
Canton	Jan 82	Jan 82	DB(4)	DB(4)	Quarterly	Quarterly
Cincinnati	Jul 87	May 87	DB(3)	DB(2)	Quarterly–99% Monthly–1%	Quarterly–99% Monthly–1%
Cleveland	Feb 87	Oct 86	IB(2)	U	Quarterly	Quarterly
Columbus	Jan 87	Jan 87	DB(6)	U	Small meters: quarterly >2" meter: monthly	Small meters: quarterly >2" meter: monthly
Dayton	Oct 87	Oct 87	DB(6)	DB(4)	Regular: quarterly Some government and industry: monthly	Regular: quarterly Some government and industry: monthly
Toledo	Jan 87	Jan 87	DB(4)	U	Residential: quarterly Commercial: monthly	Residential: quarterly Commercial: monthly
Youngstown	Oct 83	Jan 87	DB(6)	DB(6)	Quarterly	Quarterly
OKLAHOMA						
Oklahoma City	Jul 87	Jul 87	U	U	Monthly	Monthly
Tulsa	Jun 86	Jun 86	U	U	Monthly	Monthly
OREGON						
Portland	Jul 87	Jul 87	U	U	Residential: quarterly Others: monthly or quarterly, depending on usage.	Residential: quarterly Others monthly or quarterly depending on usage.
PENNSYLVANIA						
Allentown	Jan 87	Jan 87	U	U	Large: monthly Small: quarterly	Large: monthly Small: quarterly

LEGEND: DB = Declining Block; IB = Inverted Block; U = Uniform rate structure; (#) = Number of rate blocks.

	Effective Dates		Rate Structure (Number of Blocks)		Billing Cycle	
	Water	Wastewater	Water	Wastewater	Water	Wastewater
PENNSYLVANIA continued						
Lancaster	Jan 83	Jan 83	DB(3)	DB(3)	Quarterly	Quarterly
Philadephia	Jul 83	Jan 86	DB(4)	U	Quarterly	Quarterly
Pittsburgh	Jan 87	Jan 87	U except wholesale: DB(5)	U	Quarterly	Quarterly
Scranton	Apr 83	Jun 86	DB(3)	DB(4)	Large: monthly Small: quarterly	Residential: quarterly Commercial: monthly
York	Not available	Dec 86	Not available	DB(4)	Not available	Quarterly
RHODE ISLAND						
Providence	Jun 81	Mar 87: Residential Apr 87: Nonresidential	DB(2)	U	Residential: annually Commercial: quarterly	Residential: semiannual
SOUTH CAROLINA						
Charleston	Nov 86	Nov 86	DB(4)	DB(4)	Monthly	Monthly
Columbia	Sep 86	Sep 86	DB(4)	U	Monthly	Monthly
Greenville	Feb 81	Jul 87	DB(4)	U	Quarterly	Quarterly
TENNESSEE						
Chattanooga	Jul 86	Jul 87	DB(5)	DB(4)	Monthly	Monthly—90% Quarterly—10%
Johnson City	Jul 87	Jul 87	DB(8)	DB(8)	Monthly	Monthly
Knoxville	Aug 86	Jul 86	DB(4)	DB(8)	Monthly	Monthly
Memphis	Sep 86	Jan 82	DB(3)	U	Monthly	Monthly
Nashville	Jan 87	Jan 87	Residential: U Comm.: DB(3)	Residential: U Comm.: DB(3)	Monthly	Monthly
TEXAS						
Austin	Nov 87	Nov 87	U	U	Monthly	Monthly

City	Effective Date	Effective Date	Rate Structure	Rate Structure	Billing Cycle	Billing Cycle
Beaumont	Oct 86	Oct 86		U	Monthly	Monthly
Corpus Christi	Dec 84	Jul 86	U	U	Monthly	Monthly
Dallas	Oct 87	Oct 87	Residential: IB(6), Comm.: DB(6), Commercial: U – Winter IB(2) – Summer; Residential: IB(2) – Winter IB(3) – Summer;	U (Residential maximum of 40,000 gallons)	Monthly	Monthly
El Paso	Mar 85	Mar 87	IB(5)	U	Monthly	Monthly
Fort Worth	Oct 86	Oct 86	DB(3)	U	Monthly	Monthly
Houston	Aug 87	Apr 87	IB(2)	U	Monthly	Monthly
San Antonio	Sep 87	Oct 87	Residential: IB(2), Comm.: DB(2), Wholesale: U	U	Monthly	Monthly
UTAH						
Salt Lake City	Jul 84	Jan 86	U	U	Special contracts: bimonthly; All others: monthly	Special contracts: bimonthly; All others: monthly
VIRGINIA						
Norfolk	Oct 87	Jul 82	DB(2)	DB(3)	Monthly & bimonthly	Bimonthly
Richmond	Jul 87	Jul 87	DB(3)	U	Monthly	Monthly
WASHINGTON						
Seattle	Jan 84	Jan 87	U	U	Monthly & bimonthly	Monthly & bimonthly
Spokane	Jan 87	Jan 87	Residential: U, Comm.: DB(7)	U	Monthly & bimonthly	Monthly & bimonthly
Tacoma	Jan 85	Apr 87	DB(5)	U	Monthly	Monthly
WISCONSIN						
Milwaukee	Nov 82	Jan 87	DB(4)	U	Quarterly	Quarterly

LEGEND: DB = Declining Block; IB = Inverted Block; U = Uniform rate structure; (#) = Number of rate blocks.

INDEX